Song Birds of Southern India

Towards a New Paradigm of Song, Species and
Genetics of Evolution

The Author

Namratha Mogaral has a Bachelor's degree in Science with Physics, Chemistry and Maths. She has Mastered in English Literature and also has a Doctorate of Philosophy in the same discipline. She has published a book of prose fiction and Song Birds of Southern India: Towards a New Paradigm of Song, Species and Genetics of Evolution is her second book. She grew up in the coastal town of Mangalore in South Kanara district of Karnataka and part of her education was completed in Hyderabad. Currently she is Associate Professor at the Department of English of Kuvempu University in Shankaraghatta in Karnataka, India.

Song Birds of Southern India

Towards a New Paradigm of Song, Species and Genetics of Evolution

– Author –

Namratha Mogaral

2016

Regency Publications

A Division of

Astral International Pvt. Ltd.
New Delhi - 110 002

Cataloging in Publication Data--DK
Courtesy: D.K. Agencies (P) Ltd. <docinfo@dkagencies.com>

Mogaral, Namratha, author.
Song birds of Southern India : towards a new paradigm of song,
species and genetics of evolution / author, Namratha Mogaral.
pages cm

ISBN 978-93-5130-961-1 (International Edition)

1. Songbirds--India, South.   I. Title.

QL698.5.M64 2016   DDC 598.15940954   23

Published by   :   **Regency Publications**
                                         A Division of
                                         Astral International Pvt. Ltd.
                                         – ISO 9001:2008 Certified Company –
                                         4760-61/23, Ansari Road, Darya Ganj
                                         New Delhi-110 002
                                         Ph. 011-43549197, 23278134
                                         E-mail: info@astralint.com
                                         Website: www.astralint.com

Laser Typesetting   :   **Classic Computer Services, Delhi - 110 035**

Printed at   :   **Thomson Press India Limited**

....a history of the world imperfectly kept...

(Charles Spencer Darwin, On the Origin of Species, 1858)

....and she held the fiery cross of knowledge,
Eve was in paradise...

(Hebrew Bible)

Acknowledgements

I am grateful to Professor Dithi Ronen for suggesting me Konrad Lorenz's book King Solomon's Ring. Much of the journal articles came from the online journal access JSTOR for which I am grateful to Kuvemu University. I also thank Mrs Prema, Chief Librarian, Central Forest Library, Aranya Bhavan, Malleswaram, Bangalore. I also thank the publishers Astral International (P) Ltd., New Delhi for agreeing to bring out this book.

Namratha Mogaral

Preface

"What, a book of folk tales?" My urbane friend asked me when I told him that I was working on a book of birds of the Bhadra river valley. He was of course making a snide remark that fork tails (meaning birds) have become folk tales (meaning tall tales). There's a story in old Hebrew surprisingly it seems written in the coarsest dialect of the language that could have been spoken only by female slaves during the time. This story is included in the older version of the Talmud which I believe today is indecipherable and long forgotten. And since this story is written in a dialect the source of the story is greatly doubted and seldom referred to in scholarly discussions. But its neglect is perhaps because its theme is even more disconcerting.

According to the story King Solomon in his middle age discovered gold in his kingdom. But unfortunately the deposit was under the basin of a river that had the most fertile delta and covered with thick forests. The forests housed a hundreds of birds and animals in addition. But the middle aged king could not resist the temptation of some of the largest ingots of gold that lay under this lush green land and went ahead with draining the river basin. When the first cache of gold was reached the lush greenery had come to be replaced by dry desert. The birds in the forest had ceased to sing.

Now, the story goes that the king had a magical ring that gave him power to communicate with the flora and fauna. The fame of the king was spread far and wide for it was thought that his ring gave him wisdom to stay abreast of events

Figure 1: A pair of Plum Headed Parakeet

of the world. Every year the king would hold a festival to demonstrate his power to the people. Potentates and rivals from far off land would come to witness his prerogative and always went back surrendering to his superior powers. But on that fateful year, when the last fen had been cleared and the last bird song was lost the king found that he had lost his power to speak with nature. He was disgraced before the dignitaries who had come to attend his festival as usual.

Figure 2: Asian Paradise Flycather

The aging king was shaken and afraid at the turn of events. But his greatest regret was that he had lost face with his rivals and fellow men. All that the disgraced king wanted to do was hide his face from the public. Never was the festival ever held again. The king grew morose and recluse preferring to spend his days in solitude. In his heydays the proud king had trained a handmaid to keep records of his conquests for posterity. No one in the kingdom cared to remember this story of error. But the handmaid was compelled to preserve this tale. For the king in his high moment had always insisted that she record everything truthfully. How could she be unfaithful to a king's command? She wrote it down for better or worse and in the language she spoke; and has left to us who care an archetype of man's link with nature.

After all we are like King Solomon who was wise but chased after his cache of gold and lost the special gift of nature. His dilemma crystallizes the dilemma of human relation to nature: is the cache of gold more valuable or is it the magical power to communicate with nature? This book advocates the latter and is an attempt on my part to reestablish that magical power of communication.

It will seem ironical that the following account of the avifauna is derived from field observation made in the region of a dam over the Bhadra River in Lakkavalli in Karnataka, considered to be the second highest in Asia. To build the dam a good part of the local flora and fauna had to go underwater some 50 years ago. My attempts to locate printed materials recording the avifauna before the dam was built did not draw anything concrete. Perhaps something might turn up in government archives but it would require a great deal of influence to get hold of such materials. There are archeological evidences that the area forms one of the most fertile river deltas of Southern India, possibly is the cradle of some forgotten civilization. Some 30 years ago a Paleolithic (Stone Age) site was discovered with a large cache of Stone Age relics such as arrow head shaped stones and other stone hand implements. It is possible that the Western Ghats formed a series of continental reef islands or archipelago in the Stone Age, the Deccan plateau not yet being in sight then.

The tern islands are a birder's hot spot today. These islands, formed due to submerging of natural hillocks when the river Bhadra was dammed, are situated inside the Bhadra reserviour. They remain submerged in monsoon; in late winter, when the water level goes down, the islands start to show. River tern birds arrive during this time to use the emergent islands as breeding site. Up to 300 to 400 birds can be present in a season. The terns have colonized this landmass only recently, more due to the fisheries department having set up fish breeders in the locality. It is likely that the birds were disrupted from their original nesting sites nearer the coast line (since terns mainly feed on fish), when the dam drastically changed the course of the river.

Figure 3: Building of the dam must have relocated the original nesting sites of Painted Storks to further down the river course away from the Bhadra forest in Lakkavalli

Figure 4: River Tern birds on Tern Island

The dam has considerably changed the geography of the place, since it capsized more than 11, 250, 88 hectares of delta land with forests as well as cultivation land. Until some 30 years ago one would have got to hear orally circulating stories of how the colonialists, French, Dutch and British officials made trips to the delta (before the dam was built) on hearing accounts of poachards found in large numbers in the lakes and pools in winter in the countryside. Since the delta was for eons bogland habital of wild boars, it left many parts of it water logged and caused large expanses of water bodies ideal habitat for ducks, moorhens and other water birds. There is one account of a Dutch mercernary who came to shoot pochards and killed a child of six, a daughter of a lambani retinue that was camped across a lake. It is said that he gave away his rifle, the only thing of value he had on him on that day. That lambani clan is still called the rifle lambanis. This is of course a bit of local history with perhaps no official or written records anywhere and fast disappearing into the black holes of time. And it could well be possible that this book is an attempt to recapture a world gone by or fast disappearing.

The Bhadra River originates at Samse on the eastern side of the Sahyadri mountain ranges and flows southwards and east. These mountain ranges make up the better part of the Western Ghats in Karnataka that run continuously all along the western coast of the Southern Indian peninsula. These mountain ranges are greatly covered by dense moist forests and dotted with desolate villages. Kuvempu University campus that makes up a part of the territory under observation for this book is located in one such sleepy hamlet named Shankaraghatta on the windward side. To the South East some two kilometers away we are flanked by the frisky Bhadra River and its dammed serendipitous reservoir. Several lakes and watering holes dot the green countryside, ripe yellow from autumn to February with ripening golden ears of corn and paddy or mauve with the silk husks of growing tall sugar cane in December. On all other sides we are skirted by forests that became thicker as one went in deeper mainly made up of wild fig, sag, sandal tree, ashoka, oak and tall bamboos.Fruit and flowering trees abound such as flame of the forest, rain tree, jasmine, silk cotton, gul, wood apples, nutmeg, mango, jack, custard apple, guava and chickoo among others. The flowering trees wafted mysterious fragrances all the year around. The catchment

Figure 5: Spot Billed Duck

waters of the river laps against the western borders of these forest preserves of wild animals and birds, tigers, bears, jungle cats and deer, peacock and other birds in the Bhadra valley. This is the officially designated Bhadra Wildlife Sanctuary and a protected area under India's Wildlife Act (1975).

Figures 6 & 7: Malabar Squirrel & Cheetal

The location in the valley formed by the hills of the mountain range and the Bhadra River basin is idyllic. It makes up a small part of the whole Nilgiri bio reserve of Southern India. The Nilgiri bio reserve itself is considered as sharing the Indo-Malayan eco zone and forming a continuum of flora and fauna in South East Asia. Bhadra is the third largest sanctuary in Karnataka and covers some 492 sq. km area. Its forests are made up of tropical moist to dry deciduous vegetation. Not far South west some 50 kilometers is the Mandagadde bird sanctuary. Even though in recent years the numbers of birds coming to nest there has come down still the areas around it are a hot-spot for many migrating birds flying towards the watering holes of Muthodi further down the south west. The Muthodi is well known as elephant camp. Sighting of open bill, ibises, storks and egrets among others in the surrounding areas is not uncommon during late winter. Bhadra and these sanctuaries provide nesting sites for many of these birds.

The wild birds and animals from the surrounding hills and river delta use our campus as detour round the year. One was always on the look out for a cheetah or a tiger led astray by cattle or a marauding rogue elephant on the warpath. The neighing of the cheetal herd come to graze among the trees would be the only evidence that they had been in your garden under the

Figure 8: Wolly Necked Stork

moonlit night or could it have been the wild boar that nosed among the tubers you planted last monsoon? At times such as this it is not uncommon to spend sleepless nights tossing to the beat of throbbing drums kept up all night by the nearby villages to keep away tigers or panthers and other preying animals from the fattening cattle sheep, goats, and fowl. Or all night one heard the disturbing far off scolding call of a cheetal caught in a hunters trap in the jungle. These were the times of excitement and adventure for most of us here. But closer at hand on a restless sunny afternoon in spring it is quite maddening to listen to the sweet thrush that seem to be close but never quite visible. The bold and carefree laughter of the cacophonous scimitar babbler would play upon one's ear triumphantly as celebrating a savage freedom in the near by copse of trees at the edge of habitation. This was not all. On many a rainy day in early monsoon one would wake to the yearning cries of the "mia mia" of the wild peacocks. Or in late spring one could drown in the nostalgic notes of the iora's lilting melodies. And then on evenings in November there would be the noisy parrots flying back to roost after a day in a mango grove near by and the friendly "honkitovnki" of the magnificent Malabar pied hornbill.

Figures 9 & 10: Veriditer Flycatcher & Crimson Sunbird

Expert opinion has it that winter is the best time to watch birds. But strictly this is not true and seems to be more of a convention. Winter in southern India is indeed the time for migrating birds to arrive. They come from far off Himalayas and even beyond from Europe and Siberia. There were also those which visited from nearer the Western Ghats, from the Malabar region of Kerala and even from as far south as Sri Lanka. But this may be true only for strongly annually migrating large sized birds like storks. Smaller birds such as thrush and warblers usually take a detour with breaks in their migrating cycles (or migration by relay) and will persist in nearby jungles on their migration cycles across several years. This may be true for even birds such as orioles and cuckoos. Then of course there are always the resident birds. There are birds to watch all through the year.

In the past birds have been the subjects of contemplation of persons no less than the kings and emperors like Babur and Jahangir. The Muslim invader Babur left records of the birds he saw in Hindustan in his Baburnama. In the sixteenth century Jahanghir his successor devoted many pages to Indian birds in his memoirs Jahanghirnama. The colonialists were no less awestruck by the Indian avian population. Many of them pursued it as a career such as Jerdon, Linneaus, Tickell, Hume and Blyth. In fact the discipline of Ornithology developed in the nineteenth century out this colonial fascination with Indian birds. Such was its attraction.

Birds of course have always made the stock of Indian culture, art and literature. The Ramayana and Mahabharatha are replete with stories of birds and animals. The Puranas anthropomorphize birds in stories and attribute them with souls of erring men and women. Visnu Sharma's The Panchatantra is a collection of folk tales and stories on animals. They must have been circulating orally for many ages before he compiled them

Figure 11: Emerald Dove

as Panchatantra and must have made the fare of stories that every Indian child imbibed and grew up with. It is claimed that this book traveled to the west during the early renaissance and became the source for another famous book on animal stories named Aesop's Fables.

Figure 12: Forest Eagle Owl

Hugh Whistler was Salim Ali's mentor and was associated with the British Natural History Museum. Ali I believe modeled the Bombay Natural History Museum on the one in Britain. The BritishNatural history Museum houses some 1000 of stuffed birds of India contributed by colonialists, maharajahs and naturalists. In a letter Hugh Whistler makes a number of suggestions as to how to organize the birding tours in Hyderabad, Ali in 1931-32 was planning a book on the birds of Hyderabad. One advice is like this:

"Half an hour round the camp in the morning will therefore suffice to provide Henricks [Ali's bird skinner] with work on the series of common birds. Anyone who can fire a gun can produce half a dozen birds for him to get on with, pending the arrival of more important things. You will then be free to work the surrounding terrain properly to make sure that no species are overlooked. In Ladakh I used to get out early and get home about noon, and generally have a short evening turn as well. If the common stuff is dealt with by an underling at the camp you yourself can continue your own attention to bringing in more important things. If you are doing a five mile round it is waste of opportunity for you to be getting the babblers and bulbuls which are common by the camp. (Whistler in Urfi, 2005, p232)"

When Salim Ali was birding there were no laws against shooting birds (Colonial laws forbid common citizens from shooting game birds and animals such as duck and deer. Other species could be fired at on grounds of them being harmful or dangerous to humans). Any one could hire a gun and go on a spree. Times have changed now. India passed the Wild Life Preservation Act in 1975. Shooting animals or birds can get you a prison sentence even. Many Indian birds that were considered to be threatened in his days today have become nearly extinct and are rarely sighted, such as the Sarus crane, Jerdon's courser, Nilgiri wood pigeon, and the great tit. Even the birds that are still extant are steadily declining in numbers. Earlier poaching may have contributed a great deal in keeping the bird populations down but today loss of natural habitat is the main cause for the extinction of birds.

Figures 13 &14: Great Tit & Chinese Bush Warbler arrive in November

In contemporary times of course one has to rely on the camera and be content with shooting good pictures. Neither is there a use for a skinner. There has been great development in camera technology and it is easy to buy a digital hand held for a good investment. These miracle digital cameras do not need any special training and have a sonar gun which makes shooting pictures accurate. Traditional methods have been fixing a bird bath and throwing grains and stuff out for the birds. But this is not ideal for all type of species and one might end up drawing out only sparrows, crows, and perhaps turtle doves or treepies. This particular trick does not work with the grub eating and real jungle birds, such as the iora, robin, orioles, minivets, leaf birds, shrikes or even cuckoos and parrots. Professional methods such as netting and ringing are expensive affairs and require environmental ministry's clearance which is hard to come by. And they may still not serve the purpose of the research on birds in their natural setting.

The visibility of a bird species depends on their numbers in a locality. There are of course the large sized birds such as egrets, cormorants, herons, ducks, terns, and moorhens that nest in colonies. Then there are the doves, swallows, weavers, munias and bee eaters which are also colonial nesters. All of these birds regularly nest together in the same place year after year and can be sighted easily especially during breeding. Birds like hornbills on the other hand are usually found in highly visible large flocks but segregate for breeding and became scarce during this time. But there are a hundreds of other species that are found in pairs season initially or in a small sized flock post breeding. Typically there are raptors such as eagleowl, shikra, brahminy kite, serpent eagle and hawk eagle that can be found in a flock in a location what ever their general numbers. Similarly other smaller birds such as prinias, barbets, minivets, orioles, shrikes, robins, iora and the like that are usually in pairs become highly visible in flocks post breeding.

The south Indian countryside is a composite of different sorts of habitat. There are grassy open lots with short shrubs and occasional trees. These could be fallow fields let to a run intermittently. Then there are innumerable lakes and ponds or jheels that never quite dry up even at the height of a normal winter. Many of them can be found at the edge of fallow fields and open lots. They enable many sorts of birds and animals to find their eco niche. Then there are hillocks whose slopes are covered with light jungles cut through by small pathways. There are also river banks, brooks and streams. Lastly there are the jungles and forest clearings.

Figures 15a, b, c, d & e:

Long tailed orange headed lizard and other macro life

Figure 16: Forest Wagtail

On my "turns" meaning field trips covering approximately an area of 40 km radius of the Bhadra area I soon discovered that depending upon the type of terrain there were several distinct eco-systems. For instance around a medium sized pond in open grassy terrain I found birds such as wagtail, swallow, tern, cormorant, heron, starling, francolin smaller birds such as shrike, chat and bee eaters. Every pond habitat surrounded by grassy open land invariably has the same configuration of birds. Larger deeper ponds with a lot of floating vegetation and those that skirt a grove or a copse of trees have jacanas, moorhens and coot in addition to those birds around smaller ponds. I also discovered that such cache of flora and fauna occurred around any sort of perennial water source, pond, stream, tank, trough, and well or else a running tap or water pipelines.

But these terrains are transformed when the season changes, from summer to rainy, then autumn to winter. These transformations invariably reconfigure suitably the bird scenario also. The jheels during winter are of suitable depth and magnitude for certain kinds of birds such as cormorants, herons and terns. During the beginning of rainy season the same jheel becomes home to ducks and egrets. But these birds disappear to more favourable habitats when the water levels rise up as the monsoon becomes intense. Such is the case of the moorhens. A jheel with a lot of floating biomass is residence for large birds that live in large colonies such as swamphen, common moorhen, jacanas and coot. Take away the floating greenery the birds disappear. But when the water levels rise intolerably during monsoons they turn tails anyway and fly away to favourable climes.

Much of these seasonal changes are due to migrant or migrating bird population - most birds migrate at least locally – in search of more suitable habitat in each season. But change in bird scenario also takes place when certain local species of birds take over the vacated or abandoned habitat of other local species with changing seasonal vegetation. A good example for this is the case of mynas. The mynas are generally grassland birds that forage in grassy lots for grubs and flies. But in monsoon when grass lands are transformed either by agricultural activity or by nature they take over trees that are now bare of leaves and abandoned by other birds. When these tree tops were abloom in winter a few months earlier there were sunbirds at the flowers and there was no room for the mynas. The mynas have to wait for the monsoon trees to bloom to feed on nectar.

Thus these seasonal changes may be also due to inter and intra species dynamics governed by the need to maximize niche availabity. In my experience therefore I found it is important to do the watching at least for one seasonal cycle in addition to doing the different terrains of a location. Moreover the connection of birds with

Figure 17: Red Spurfowl is a forest ground dweller

a terrain is combined with the food they eat. Ducks inhabit jheels that do not harbor certain kinds of fish. But on the other hand availability of food can change with the season.

Figures18a, b, c, d, e & f: Some flowering and fruiting trees

Perhaps it is also important to know which tree or shrub is associated with what bird. Not only a variety of birds inhabit different trees, they also forage at different levels of foliage. As a rule they prefer trees with foliage the same size as they. For instance cuckoos are found on custard apple or mango trees. The iora prefers sandal and wild jackfruit. Owls inhabit broad leaved groves. Bee eaters when not foraging the air, sit on bright shrubs and wintering trees in open grassland. Hornbills are frequently found on wild fig trees. Many others such as lark, shrike, munia even pitta and fowls of various kinds are found on the ground in the grass or in low lying shrubs and weeds. The raptors typically keep circling in the sky, but when the flock is together they do sit on tall trees tops. Then there are the water birds ducks, mallards and others always in large coteries.

Figure 19: Blue Tailed Bee-eater

Birds even the smaller ones can migrate to a long distance. But it is more likely that they are local migrants flying seasonally about 100 to 300 kilometers, *e.g.,* cuckoos, thrush, and woodpeckers. Such birds are attached to their places of birth and will always come back seasonally. They seem to have a prerogative over the place of nativity, where the non avian residents and even humans are inducted into the routine of bird life. Long distance migrants, like Eurasian eagle owl and Eurasian golden oriole can become assimilated locally. Birds of a particular area tend to conserve the resident configuration and do not alter it for any reason. Smaller birds are often

Figure 20: Oriental White Eye

Figures 21a & b: Rufous tailed & Brahminy starlings found in South India are long distance migrating

transported in between larger birds that are flying in a flock, *e.g.,* oriental white eye and warblers. Larger birds can forcefully carry them to maintain the configuration in an area.

This book is devoted to the study of song birds. In ornithological parlance an order of about 4000 species of birds named as oscines or passers classified under Passeriformes are attributed with the capacity to sing. The song birds I have undertaken to study are not all passerines. Among those that may be viewed as passerines are the sunbirds, iora, tailor bird, thrush, wren warbler, magpie robin, flycatcher, leafbird, bulbul, scimitar babbler, tree pie and oriole. In addition I have also included the barbets (piciformes) cuckoos (cucliformes) and doves (columbiformes). These are the song birds of South India and all of them merit the label song birds. Any orchard or forest edge here is filled with the various melodies of these birds that sing all the year round throughout daylight hours regardless of season. There are many others like the flower peckers, prinias, munias, bee eaters, shrikes and wagtails that make sweet sounds but these are the sweetest performers.

There isn't much research or writing on the song birds of India. In addition because birds are a tricky subject to observe being full of camouflage and concealment most books on bird song circulate a lot of wrong conceptions. Books that are written on the basis of laboratory experiments will never give the true nature of birds because they are based on study of birds in captivity. The observations presented in the following pages are worth the while for they are derived from watching birds in their natural everyday settings without the mediation of laboratory equipment. Chances are that these very same birds are becoming rarer as we transform the ecology for human needs with more and more elaborate technological measures. And my book could well be the last line on a world fast disappearing.

Namratha Mogaral

Contents

List of Photo Illustrations, Tables with their Captions

Preface

Chapter One   •

Chapter Two

varying from one line to five and seven lines. The first stanza made up the body of the song narrative with imitative references to one or many other bird species in progressive order in the territory followed by a second concluding stanza sometimes with a one- line conclusion but always with a pattern of notes comparable to a word list in children's school language book. It may be conjectured that the member picked up his song routine from a school in his territory

Chapter Three

Appendix 1

1

Understanding Bird Song

"....And learn to chant a tongue men do not know...."

(A Rose upon the Rood of Time, W B Yeats)

What makes the birds to sing? The nature of bird song has intrigued ornithologists for long. The earliest enquiries lead to an important discovery; birds that sing have a remarkably higher level of the hormones testosterone and oestrogen in the body. So they attributed bird song to the ability to produce the hormones. One hypothesis is birds are genetically programmed to sing and therefore sing out of instinct. This would mean they have an internal biological protocol that triggers hormones and tells them when to sing and how. It is possible that this internal protocol is initiated by nothing more than the good climate or perhaps certain foods. Today there is no doubt that birds have a hormone controlled nueral or brain mechanism for producing and recognizing songs. Experiments conducted on captive birds during fledging time (mainly conducted on a Eurasian starling fledgling by Ziegler & Marler, 2012)) when they are just learning to vocalize and the study of the brain during this stage with the help of brain scanning (EEG mapping) have provided ample proof that birds have brain localized song control system (see

Picture 1: Common or Eurasian starling has a range across Europe and Asia. In India it is found in the Himalayas (photo by Pierre Selim, sturnus Vulgaris (France), 2012-02-26-2jpg, Wikimedia creative commons)

Figure One below; Brenowitz 1991; Brenowitz *et al.*, 1999). This system involves several regions of the bird's brain. Certain proteins (mainly a protein named ZENK; Jin & Clayton 1997; Mello & Ribeiro, 1998; Jarvis, *et al.*, 2000) released by the brain cells when activated by external and internal stimuli wire these regions into a system and to the bird's syrinx (vocal chords). The development of the control mechanism is found to be related to the levels of two sex hormones testosterone and oestrogen found both among young females as well as males (De Voogd, J J (1991); Kirn A P

& J J De voogd (1999)). In other words this would mean they have a sex controlled genetically adapted song mechanism.

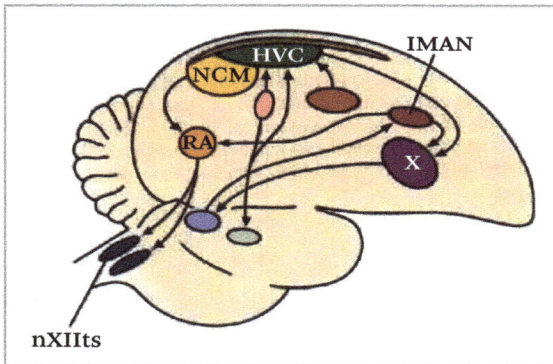

Figure 1: The song system of a typical song bird

The major components, or nuclei, involved in song production include the robust nucleus of the arcopallium (TRA), the higher vocal centre (HVC), the lateral portion of the magno-cellular nucleus of the anterior nidopallium (IMAM), the caudomedial neostriatum (NCM), and area X(X). Neural pathways carry signals from HVC to nXIIts to the muscles of the song producing syrinx. Other pathways connect the nuclei such as IMAN and area X, that are involvd in song leaning rather than song production (after Brenowitz *et al.*, (1999) as cited in Alcock (2005).

More recently field study of songs of birds has foregrounded a second hypothesis, which claims that bird song is a kind of social behaviour in response to the presence of other birds, animals and phenomenon in the environment (Armstrong, 1973; Payne, 1981; Baptiste, 1996). This second explanation resembles our understanding of human capacity for language. Theories of human language have always emphasized upon the social dimensions of linguistic behavior of humans. Likewise songs of birds must indicate their gregarious nature. They are the index to the social interactions among birds. But mere field observation does not become a theory. An experiment was conducted to test the hypothesis. A zebra finch was raised in isolation in order to find out if it developed song like zebra finches in the wild do. The isolated bird produced exactly the same song as the members in the wild but its songs were interspersed with long abnormal silences. This isolated bird was further allowed to intermingle with another fledgling that had yet to learn its song. It was found that the new member produced a song closer to the wild type than its tutor. In about four generations of song learning the isolated zebra finches had got back their wild song. For the scientists this experiment proved two things, one that bird song is primarily a genetic ability and the other that it has a social dimension but of a limited kind because the finches did not produce any song beyond the fixed number of notes they produce in the open wild (Fehr & Techernichoviski in Bolhuis & Everaert (eds). 2013; see also Thorpe, 1958 on song learning in chaffinch). Thus today it is held that bird repertoire for producing social meaning by sounds is far smaller. While human language is considered to be predominantly social in character bird vocalizations are thought to merely typical produced more by biological instinct. Most experts today prefer to combine the two perspectives and find a socio biological understanding of bird song.

The above experiment throws a great deal of light on bird capacity for vocalization than scientists are willing to allow. Zebra finches are actually flocking birds that in the wild always exist in large sized flocks of upto twenty or more. In addition they are not born singly but in a good sized clutch. The isolate must have been separated from such a clutch. The abnormal silences in the isolate song did not reflect any abnormaility in the song production but merely expressed his social isolation; it communicated the absence of fellow members and therefore his song production was at the outset social in nature and meaningful. Even in the natural setting most birds actually string together sounds that usualy refer to (or address to) members in the territory. Song length and increased number of notes can indicate even the intergenerational continuity among members. Members not related genetically may be referred to by imitating their vocalizations; kins are addressed by the species typical notes. A silent member who cannot sing in the territory can be represented as a silence in the song. Thus in the wild the length of a song is a good indicator of the composition of a local bird community and the hospitable ecological conditions. That's why in the experiment when the isolation was decreased the members produced the wild type song in full. The isolates song merely reveals the context-content dependency of song pattern of finches and gives a clue to the pragmatic aspects of bird song communication. It also places before us the possibility of interpreting bird song as made up of discrete sounds each with fixed semantic range as in human langauge. This would give a new direction from the predominantly structural view of birdsong.

Further, if we examine the experimental results and compare it with observations on language learning among isolated humans (for e. g the wolf boy) it would appear that humans acquire language by sheer imitation rather than having it genetically, while bird songs are a genetic ability but that both humans and birds produce vocalizations for the purpose of communication that involves sociabilty. We know the nineteenth century story of the wolf boy (immortalized by Rudyard Kiplings as Mowgli in The Jungle Book) who grew up amidst animals in the forest and learnt to communicate like them. When he was brought back to human society he tragically failed to learn any human language and became an embarresment to the people and had to be shot. It would be wrong to view the finches' song as less social than humans'. It would be more appropriate to say that linguistic behavior must be innately social in all communicating organisms. Niether would it be correct to conclude that the bird song is less evolved rather there seems to be greater parsimony if we consider that it can communicate as many social meanings as human language.

Communication among birds is an elaborate process and is not limited to vocalization signals only. It is rather a complex semiotics of vocalizations, bodily movements such as wings beats and postures. Touch, smell and colour coding also play a part in this semiotics. Therefore even though it may be insufficient to limit communication among birds to vocalizations, as in human vocalizations seems to dominate that process. This may be claimed especially about birds, reptiles, fishes, ceolentrata and other kinds of insect, butterflies *etc.*; on the other hand touch and smell seems to dominate non human placental animal communication, for example cat, dogs, squirrels, *etc.*

Pictures 2a, b, c, d, e, f, g, h & i: Poses of birds

Organisms that produce vocalizations are equipped with special physical organs for the purpose. Both birds and humans have similar vocalizing organs made up of three parts, vocal chords or syrinx (known as larynx in humans), vocal resonator or trachea and organ for articulation or mouth and its parts. The syrinx is the place where the sounds orginate while the resonator and the mouth parts shape and modulate them. The most striking anatomical feature of the bird syrinx is that all the sound producing components in it are present in well formed duplicates even though it has only one trachea resonator. These distinctive double vocal chords are wired to the brain separately and can be used separately so that a bird can sing in two voices. Analysis of bird song has revealed that usually it is composed of two harmonically unrelated sound patterns simultaneously (see Nowicki & Marler (1988) for an account). But in addition there can be a third element composed of mixed frequencies. This perhaps explains the voice throwing (ventriloquist) ability of birds. The two sound boxes with one trachea resonator produce a time lag between the two voices resulting in a stereo effect. This enables the bird to conceal its real location and make it appear as if the sound is coming from many places at the same without ever changing its physical location.

The discovery of the duplicated vocal folds of the bird's syrinx which are the sources of its two voiced vocalization lead scientists to discard comparison of bird songs to musical instruments (Baptiste, *et al.*, 2005). A model more comparable to humans was proposed. In a wind musical instrument there is no syrinx but only the resonator. Still scientists have been puzzled over the fact how the two vocal chords are used with only one resonating trachea tube (Greenwalt, 1968). It may be surmised that the length of this tube (or birds can retract their neck and shorten or lengthen it) is far less in proportion to the sound sources (the acoustic membranes or vocal folds) and birds continue to produce sounds like humans (currently it is thought that in humans sounds are produced by a single vocal chords and then is modulated by the trachea by muscles as it proceeds towards the mouth). So that not only can birds produce two voices they can also modulate these two voices so as to produce a third voice (Nowicki & Capranica 1986; Nowicki, 1987). Scientists today conclude that bird songs (single frequency sounds are produced) are comparable to human singing while other sounds like birdcalls (sounds with varying frequencies are produced) are akin to human's talking (Nowicki, 1987).

Yet Ornithologists doing field study on bird calls and songs have so far failed to come to a consensus on the actual sounds involved in each bird song or call. They have found it near impossible to go by the human ear. For this reason research on bird song is done with the help of gadgets.   Studies of bird songs carried out with the help of recording devices depend on the electronic data produced by a sonogram or sonoscope that convert the actual songs into digital data and imaging. By Analyzing these spectrograms or sonographs of song patterns two basic types of notes or sounds are typified; whistle (identified as horizontal patterns in a sonograph) and trill (and half trill; identified as vertical bars in a sonograph;) is also considered to be the repetitive element in bird song; see Figures 2a & c). This is labeled as a bigram pattern. Bird language is thought to be entirely composed of sequences of whistle and trill, what Berwick *et al.*, (2011) term as "strictly locally 2-testable languages' a highly restricted subset of class of regular languages". That

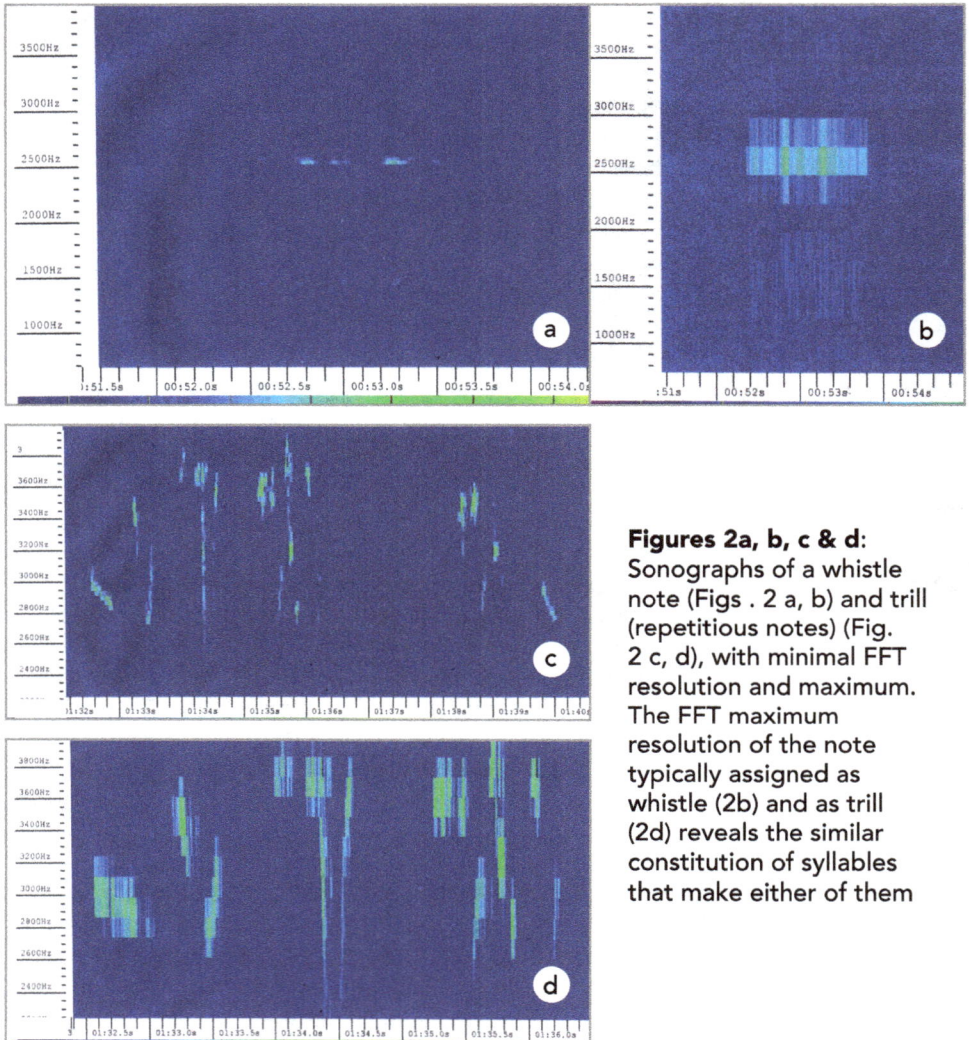

Figures 2a, b, c & d: Sonographs of a whistle note (Figs . 2 a, b) and trill (repetitious notes) (Fig. 2 c, d), with minimal FFT resolution and maximum. The FFT maximum resolution of the note typically assigned as whistle (2b) and as trill (2d) reveals the similar constitution of syllables that make either of them

is to say it does not require a great deal of proving and that it may be truistically claimed that most birds do not produce more than two types of sound patterns and therefore may not be considered on par with human languages. Actually research on the structure of bird phonology couldn't be more anthropocentric and short sighted than this. If we look at the so called complex langauge of human it is evident that it is also made of two sound types, vowel sounds and consonant sounds. The entire human language is the sequencing of these two fundamental sound types. I suggest we must recognize that what is called as whistle in a bird song is composed of a consonant (or consonant cluster) followed by a long vowel (diphthong of human language) and/or a semivowel; and a trill is composed of one or more pairs consonant (or a consonant cluster) and short vowel combination. This is also the basic pattern of the syllables of human phonology; consonants are always combined with vowels in actual speech production. In fact this is the universal and logical feature of vocalizations. The horizontal patterns visible in a unresolved sonogram and identitfied as whistle is thus produced by the long vowels and verticle patterns identified (when the sonogram is minimally resolved) as trill corresponds to a variety of consonants or consonant clusters of human language (see Figure 4c).

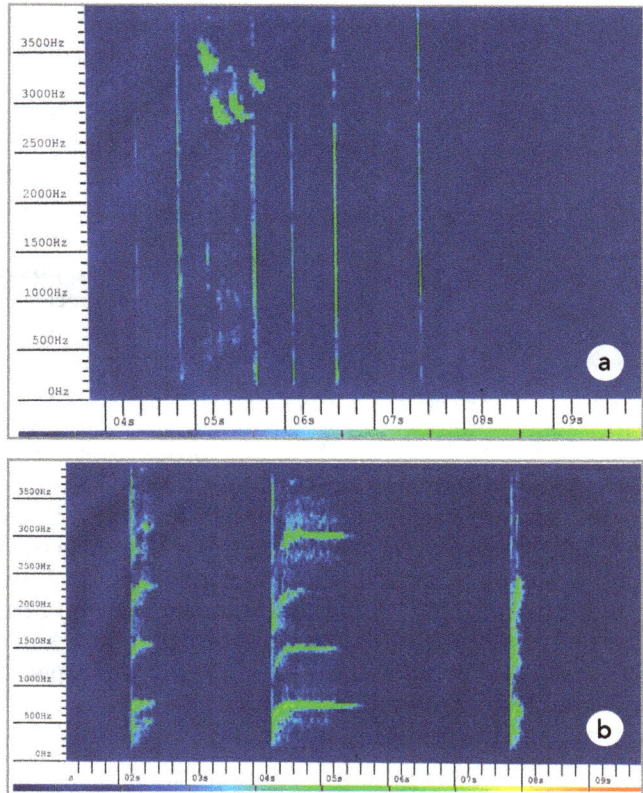

Figures 3a & b: Reveal that bird song is produced at higher frequency note initially (Fig. 3a), while human sound is produced at lower frequency word initially (Fig. 3b). Note also doubling effect of continuous vocalization occurs in the higher frequency range for humans and in the lower for birds. The main reason for this could be that bird vocalization is predominantly composed by ingressive sounds

Indeed the sound patterns and the syllables of calls of birds can be unlike that of humans to the ear. For example "twe" sound can sound like a whistle as in the Tickell's blue flycatcher's song. The "r" sounds like a series of multiple rolls in the "tweeter" of bulbuls or "chirr" of warblers. Similarly the "cu" sound is velar in the "cuk" of cuckoos, and the "ee" vowel sound in the crimson backed sunbird's song "tweenfreen" ends with a nasalized "n" sound. You cannot fail to appreciate the thick billed flower pecker's distinctive "twieny" sunbird's "cheep" or the baya's "tweentween". These sounds all seem to be far from sounds of human language. Birds seem to have a language all of their own. Bird's produce vocalizations that are predominantly ingressive (sound is produced with pulmonary inspiration). This must explain the rather horn like shape of the mouth channel *i.e.*, the place of articulation. Birds do not have lips and the overall shape of the mouth with the beak must be different. Their teeth does not include molars (they use stones kept in the gullet to grind hard foods) and the canines and incisors make up an even saw tooth lining of the front inner side periphery of the beak. This must account for song notes that hear like whistles (or for the sliding of consonants) and for the predominance of back velars and glottals, retroflexes or multiple rolls, nasals and interdentals. Actually there are many human languages with predominance of the latter sounds, such as swahlili and other older African languages or Chinese and Japanese and Sindhi which has ingressive syllables.

Picture 3: Little Flower Pecker

Analysis of a broad cross section of vocalization data of birds points to some interesting features. Most importantly, birds produce highly structured, inflectional language, *i.e.,* one can identify distinctive lexical stubs that take on inflexes to produce new meanings (combinational rules). Secondly, it also suggests that avis as a class possess a collective language, which is accessed by individual species in fixed ways based on their genetics. That is to say each species has claims to only a part of the avian language and strong rules of linguistic exclusion govern the actual production of vocalization. For example, at least two different groups of birds may be identified in the Bhadra region, the twee-s (magpie, Tickell's, iora bulbul, leafbird *etc.*) and the kra-s (treepie, oriole, malakoha, parrots, *etc.*). The twee-s never produce the sounds of the kra-s because they observe a linguistic taboo. This linguistic exclusion seems to be governed by the species place in the avis evolutionary tree and can give information about the species formation. Birds thus may be said to use language magically, as secret or cryptic codes. Breeding time vocalizations are especially of a magical nature and communicate by encryption, like a riddle. This may have been an aspect of human language in the primitive times but not any more.

The inflectional language of the avis class suggests that like human language bird songs involve naming and classing of things and objects, drawing relations, and expressing internal states of being. It does not seem farfetched to claim that birds have something close to meaningful or lexical units. These same lexical units seem to be produced across species albeit in different dialects. For instance a basic unit is the sex differentiating and differentiating between adults and fledglings repeatedly produced across a population. For instance "krya" as in the rufous treepie's "kukra krya ku" indicates that it is a male bird. The sound "ma/mya" is associated with female as in for example female tree pie's "kukra mya ku" or shikara female's in-breeding "kiiowmiiow,miiow kiiow" and "miaw" of the breeding peahen both of which sound like the mewling of a kitten. The sound "pip" in the "draan pip" vocalization of the king drongo male in his first season could well refer to his status as juvenile. It seems to be distinctly possible that bird songs transmit information about plumage colour and pattern as for instance in the male purple rumped sun bird's "chleep cheleep"referring to the shades of colour wash that flash on the wings of this bird.

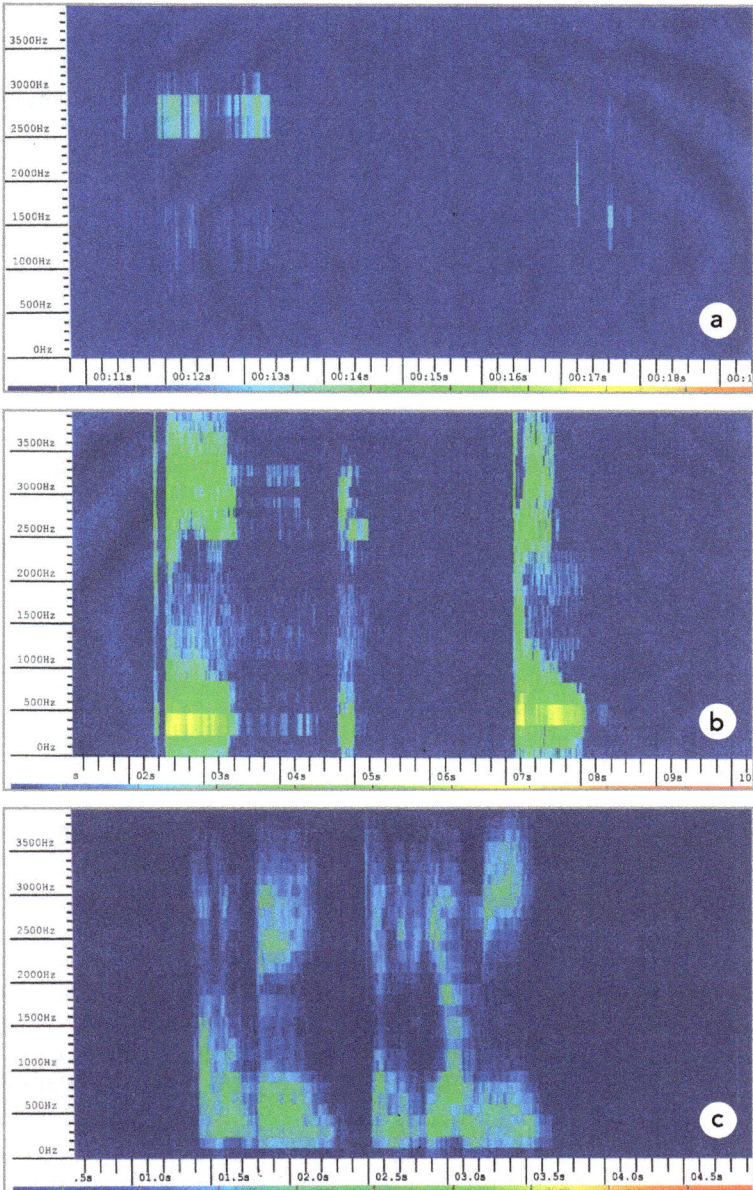

Figures 4a, b & c: A spectrogram for the "tweeee, tweun" pattern in the iora song (Fig. 4a) is nearly identical to the spectrogram of the same sound voiced by a human (Fig. 4b). Fig. 4c shows the frequency resolution for four human notes (words) in connected speech pattern "hello there, cn'u hea'me`" Note how consonant-short vowel pairs that make up a word are shown vertically (trill) and lengthening of vowel sounds are shown horizontally (whistle)

Table 1: A possible conceptual grammar of avian communication

Song units	Lexical value	Species
Kra/kria	male	Treepie, crow, oriole
Ma/mia	female	treepie
Miow /miaw	predatory breeding female	peafowl, shikra
Kaiyey	sister	hornbill
Pip	juvenile	drongo
Ka/ku/co	to sing (sex differentiated)	treepie, malakoha, dove
Kru.krau,kuruwi	singing fowl	Oriole,
co/cu/ku	singing fowl/ fowl at large	dove, barbet
ki/ kiki/ kiiow	predatory bird	shikra, hornbill
kukru/ukru	incubating	treepie,oriole
tavi	tree hole nester	robin
cuk/kkh	hidden nest dweller	barbet, coucal
tava	not tree hole nester	nilgiri flycatcher
tuwi	leaf nester	tailorbird,minivet
twe	weaver	robin, tickell's
twi	fine weaver	jungle prinia, whiteeye
tchu	spittle nester	sunbird, grey brest.prinia
hornki	with horn	hornbill
che	with shiny plumage	sunbird
pe	small bird	sunbird,robin
tchewee	with stylish beak	sunbird
wee	wonderful	leafbird
yan	I	drongo
droin/druin/doin	moist or dewy forest dweller	drongo
troin	tree hole nester	drongo
eny.ani	several	flowerpecker,
-eru/erui/ik/ip	be	sunbird,malakoha, drongo
-irah/ -a	not be	parrot
-yaru/aru	of	oriole
-t, -k	circumscribed or contained	bulbul,sunbird

Most birds produce different vocalizations at different stages of their life time that seems to code information of the bird's status especially in connection with reproduction. A bird that has contributed to the reproduction of chicks in any way, announces of this to the entire community by producing sounds that communicate this information. It is also possible that there are distinctive vocal units that are repeated across bird species that refer to type of nesting such as "tavi," "tava" "tuwi"and "twe" tree hole, not tree hole, leaf nesters and fibre weavers, respectively. As for instance in the Malabar pied hornbill's "hornkitavnki" the "tavan" refers to tree hole nest. Similarly "twieny" of the flower pecker means the birds weave young short fibres without much turning. "Tweentween" of the baya weaver refers to using long green fibres turned several times. It appears that songs transmit information about the nesting habit of a bird and the way the nest is built, the materials used *etc.* They announce the bird's status niche in his habitat. Sound patterns also indicate the genetic makeup of a bird such as "co or cu" meaning singing bird, "ki or kiki"predatory bird. Songs can transmit other kinds of elaborate social meanings:

The Tickell's blue flycatcher for instances produces different songs at different times of the day Thus when he sings his songs at different times of the day it could mean different things but transmitting his response to his surrounding.

Ornithologists are of course reticent to assign such human-like expressive capacity to birds and other organisms. Studies of song learning among birds such as song sparrow report that the learning invovles merely structural mimicry and memory. Similarly an experiment carried on parrots indicated that the learning did not actually include understanding the meaning but rather structural memory (Marler & Mundinger, 1971; Baptiste, 1996). Even though this could be merely a situation comparable to that between birds and us humans, after all we can produce

Picture 4: Tickell's Blue Flycatcher   **Picture 5**: Rose Ringed Parakeet

many meaningful languages but we cannot understand a bird song but as data when graphed. Still I must here also add that a big stride has been made towards a comparison of bird song with human language based on molecular research. For instance it is fairly well established that vocalization capacity among humans and birds are related to the same genes FOXP2 (for vocal learning), MCPH1 and MCPH5 (for tonal langauge) in these organisms (see Fitch & Mietchen in Bolhuis & Everaert (Eds). 2013). This is of course hypothesized on the basis of in situ observations of gene expressions in the two organisms during actual song learning and recognition. It is even speculated that birds and humans may have a common ancestry and that this is not merely a case of convergent adaptation.

Similarly it is also remarkable that human language learning involves the structural patterns of the language comparable to the structural mimicry among birds (see Moorman & Bolhius in Bolhuis & Everaet (Eds), 2013). A third clue is the generalizing ability in vocal learning that birds' display, that means they can recognize a sound pattern even when it is produced in different contexts and by different members. They can also recognize similar sounds that fall in the same acoustic range (Ryan *et al.*, 2003). Currently it is thought that these similarities between human and bird language are merely behavioural (Berwick *et al.*, 2011).

Thus it still remains to prove how birds relate these patterns to meanings like humans do in language use and we dodgedly continue to study bird songs as quantities of meaningless sonic data.

Birds display highly developed uses of their songs all the time: The crow a common bird in India uses the song to eco localize its surroundings. For instance crows always craw as they start to fly or while reaching a perch. This crawing is made up of the same note produced at several different scales almost simultaneously. This vocalization differs a great deal from the much softer and shorter vocalization produced by nesting members. Songs can be used to transfix a prey just a barrage of words may be used by humans to overcome dissent. A pond heron uses a sub decibel siren-like "ki" vocalization to transfix its prey like this. This is somewhat similar to mobbing calls used by predatorial birds before they attack the prey. But species may differ in the mobbing calls they produce; for instance the white cheeked barbet does not sound a bit like the Brahmny kite. A good song can be used to distract a predator just as cryptic plumage can. Birds like thrush, lark and chats produce stunning songs which are meant only to divert your attention what I have called as decoy singing in this book. Leaf birds use mimicry accompanied with ventriloquism for the same purpose. Some other birds, *e.g.,* tailor bird, use alarm calls (see Figures 6a&b below) to warn other members of a possible threat in its habitat (see section on tailor bird).

Picture 6: Indian Pond Heron

Songs of birds can be broadly categorized as functional or/and social. But it may not be possible to use these categories in mutually exclusive or in an absolute way. Some kinds of vocalizations may be more functional than social while others more social than functional. In ornithological literature two kinds of bird songs are recognized: identity songs and territorial songs (Marler, 1957; Krebs *et al.*, 1978, 1981; Kroodsma &Byers 1991; Kroodsma in Kroodsma & Miller, 1982; Stoddard in Kroodsma & Miller 1996). This classification is based on the social uses songs are put to by the birds. Both these songs are attributed to the breeding male. Songs that transmit information about plumage pattern, reproductive status, nesting and feeding habit and genetic information that I elaborated upon in the previous page (p8) may be called as identity songs. The second kind of song, territorial song is used to demarcate and monitor the territory. In actual field observation you will discover that birds do not have different songs for identity and territory but may use the same song for the different purpose. The territorial song of many birds includes their own identity songs as well as the mimicry of songs of other birds present in the territory. Whether such a territorial song is directed at building a friendly relation with fellow members or rather an aggressive assertion of territorial occupation as is usually attributed to territorial songs is a matter of speculation. One will also notice that different members of a species flock take to different kinds of singing depending on their role and status in their flock. For instance, among the orange headed thrush a female bird undertakes decoy singing while a male does the identity singing. Among the Tickell's blue flycatcher the males do territorial songs during the day while the breeding female monitors the territory in the nighttime by producing nighttime singing. Among purple rumped sunbirds the male safeguards territory with song during off season, the female takes over during breeding season. Among some species such as magpie robin identity song is produced by the breeding male and territorial song by the surrogating male.

I have already said that vocalizations among birds are accompanied with bodily postures and wing gestures. All birds have well established song routines or ritualized delivery of song that are transmitted across generations: Songs are delivered at fixed times of day and season and also suitably to the occasion. In addition they have established modes of delivery which involve a fixed way of holding the wings, beak and tail. The lack of diversity of sounds is made up by the sophisticated use of intonal patterns and other body language. Very subtle meanings are communicated by simple repetition of the same sounds to different rhythms and tonal variations, producing song pattern variations accompanied by body poses. In addition the song routines are usually delivered from preferential locations in the bird's territory, some times even involving preference for fixed levels on trees.

If you listen to the birds around you it is not long before you find out that birds make several different kinds of sounds. In bird literature these vocalizations are

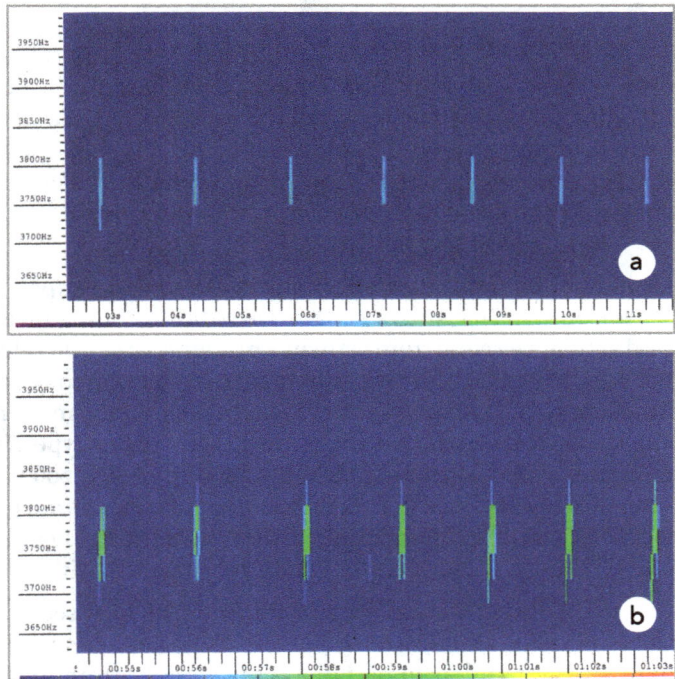

Figures 5a & b: Spectrograms of the female Tickell's blue flycatcher's nighttime territory monitoring songs on two consecutive nights composed of a note each wee & twee. Subsequent songs combine these two notes in sequence such as twee, wee, twee. (see Figure 4 of chapter Two for the male's territory song)

grouped separately as calls (innate calls, Marten *et al.*, in monograph series 2011, p. 71) and songs. One apparent distinction that is made between calls and songs is that calls are usually carried on in a call and response mode between a breeding pair and in many cases including the chicks while songs are usually sung solo. Take the example of lapwings. In India these birds inhabit open grasslands almost throughout the year and are found either in pair or in medium to large flocks. Upon sighting a possible intruder the leading bird among them starts to call out what sounds like this "athadaavate, thathadaavathe" repeatedly and finally takes off on wing to alight a little further in the field. The other bird in the pair follows suit crying out in reply "thap thap." It is evident that these calls are used to communicate between a leading pair and direct the flock to act. This sort of communication may be found even among the babblers, a large family of passeres all the members of which move about in small talkative parties of parent and fledglings all the year through. Among the house sparrows the leading bird emits an almost human sounding whistle to hustle members of its flock together. Members are prohibited from movement outside the monitoring bird's whistling range. Contrast this with the scimitar babbler's individualist scolding vocalizations. They are never found in any large numbers and mostly singly. They make several different types of solo vocalizations all of them which are patterned and sonorous. The in-breeding male's song are heard

Picture 7: Red Wattled Lapwing

only in season *i.e.,* monsoon. On the other hand the scolding outbursts of the pre-breeding female can be heard all the time. Both these vocalizations are apparently made without a particular address and do not seek a response from other birds. Similarly, the thrush or the iora in an orchard in Southern India, their songs are sung solo and seem to be directed at the whole territory.

A second apparent distinction made between calls and songs is that calls are thought to be routine vocalizations while songs are specifically linked to breeding season only. In fact, singing is thought to be the prerogative of the male and breeding bird at that. Deriving from Darwin's principle of sexual selection it is concluded that the song of a male bird gives sufficient information to the prospective female on his fitness to breed and plays a role in mate selection. A more recent formulation of the above classical view is that songs are thought to be a type of specific mate recognition mechanism (SMR System, H E H Paterson in Mallet, 2007, p. 430). This theory about song of birds assumes the sex distribution of songs and reproductive functionality. In fact all bi parenting animals, where the breeding pairs need to communicate with each other for the purpose of reproduction are attributed with the SMRS.

It is important to keep in mind that songs seem not to be different from calls in kind, but only in degree. And that they are social addresses to other birds in the habitat and form part of a semiotics of communication. For one thing both calls and songs need not have a specific address or specific response. It is possible that these very same calls develop into elaborate songs as a season proceeds. Non passerines such as storks, cranes, swamp hen and water hen that can barely cluck produce elaborate call & response as part of mating rituals. Passerines such as robin, warbler and prinia develop their calls to fully fledged songs often sung from tree tops addressed to the entire community. Among hornbills of the Western Ghats, calls make up a part of a mating game. Season initially (June for Indian Grey and November for Malabar) a gang of eight to ten young hornbill pairs flit from tree to tree chasing and calling out to each other "kiiiii" flapping the wings like hands. This group ritual seems to be crucial for deciding among the breeding pair(s) for the season but not for the pairing between the sexes.

First of all it must be noted that field observations suggest that song distribution between the sexes is governed by some kind of biological economy relating to type and extent of reproductive participation. This biological economy can take different forms and result in varied song sex distributions. This gives scope for considering that sex song distribution has less to do with sexual difference and more with reproductive participation. Among the non passerines that produce elaborate vocalizations the coppersmith barbet (piciformes) breeding male can sing at the start of the breeding season. But as the season carries on he falls behind in song like his female. Not so among the white cheeked barbet the male continues to sing for

Picture 8: Malabar Whistling Thrush male is famous for its well developed song notes

the better part of the year. This can be traced to the differences in the incubation method: the coppersmith barbet has to sit on the eggs but the white cheeked simply seals up her tree hole nest during incubation. Male doves (columbiformes) are as silent as their females during the incubation period because during this time they have to feed their sitting females. Watch the shikra (falconiformes): the male is vocal in initiating the breeding season but soon after that the female takes over. Her in-breeding calls are even more raucous than the male's he sounds juvenile in comparison. In both these cases it seems to suggest that singing capacity depends on the extent and type of participation in the actual reproductive process. Among the shikras the female takes the lead in providing for the chicks and the baby-sitting male. The male does not produce any call in his sedentary role as resident guardian of the chicks.

The passerines provide even more curious evidence: among them it is a matter of choosing between singing or breeding, that is to say there is greater likelihood for a bird that opts to breed falls behind in singing regardless of the sex. Among the

Picture 9: A pair of White Cheeked Barbet on a sandal tree

Tickell's blue flycatcher the female can sing the species typical song of the males but only up till the time she lays her eggs. Breeding males among ioras for instance cannot produce the species typical song at any time of their life. Instead if in their first season they produce a poor imitation of it and that too only at the start of the season, comparable to that of their female's weak song. Among the magpie robins the breeding male cannot produce the more elaborate song of the singing male who is in the role of surrogate parent. Thus it is not true that songs are distributed across sexual differences. It is more likely there is an inverse relation between quantity of singing and length of reproductive involvement whatever the sex of the bird.

Secondly, among many species especially passerines sustained singing can be had at the cost of a complete sacrifice of the reproductive privileges. The singing bird seems to enter into a kind of breeding coalition with her/his fellow members where (s)he plays the role of singer and relinquishes the breeding role to another member. Intriguing fact is singing enables the bird to establish domination that reproductive functionality does for other members in her/his flock: But this can be understood only in terms of social dynamics. A most clinching evidence in this regard is found among iora, magpie robin and tailor bird. Among the tailor bird the singing bird is always a female. The breeding male and female both do not produce any song. And comparable to the singing males of the iora birds and robins she is unable to produce any viable eggs and is inducted into a reproductive coalition as we shall investigate in this book. Therefore it is important to see that significance of song must go beyond pair breeding. Song must have a function more than that explained by the principle of sexual selection or as specific mate recognition mechanism. Only if we uncover this secret will we understand song distribution among male, female as well as among the members of various bird species.

In the natural setting different bird species produce different sound patterns or song types. This descriptive classification is based not on the uses of bird song (identity, territorial *etc.* I wrote of earlier) but on their syllabic composition (song notes) and mode of delivery. A song type may be distinctive or unique by differences in syllabic pattern or notes (a note is used to refer to syllable clusters separated from the next note by silence in a song), tone variations and/ or mode of delivery (Searcy, *et al.*, 1999; Marten *et al.*, in Renner & Rappole, 2011). Analysis of song is usually done by identifying the minmal units of production (MUP), *i.e.,* unique clusters of notes that make up a song, some of which may be repeated in it (see Podos *et al.*,1997) with the help of spectrograms of the sound data. Such a quantitative analysis of

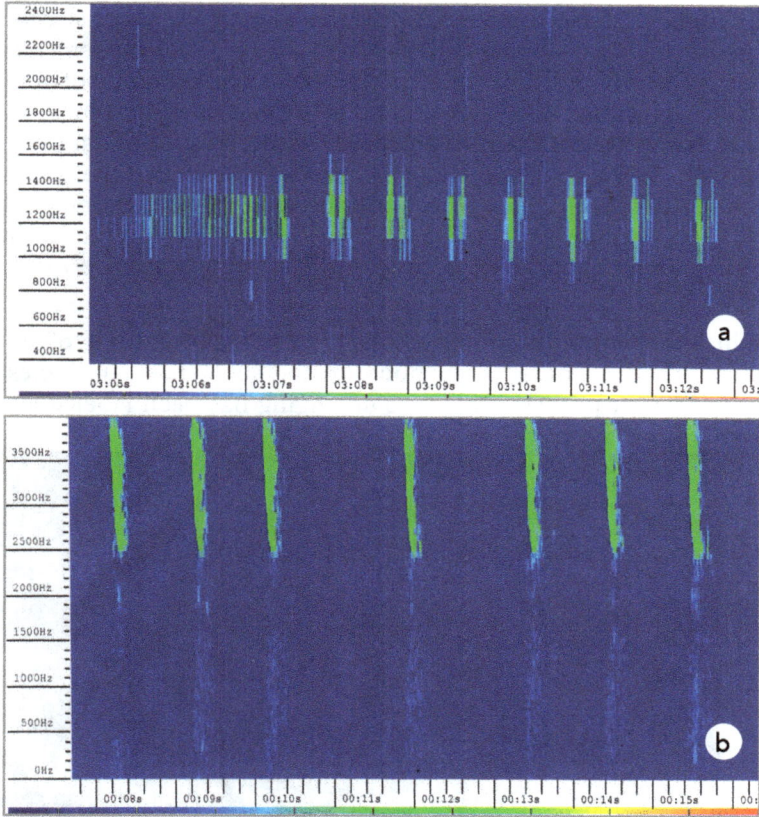

Figures 6a & b: A Mobbing call of the white cheeked barbet (male). The second spectrogram shows alarm calls of a female tailor bird

songs has lead researchers to conclude that a bird can have a song repertoire (a collection of songs) with up to seven to thirty songs or types (Kroodosma, 1982). Strictly speaking the concept of MUPs and assignation of song types with their help is a fiction of methodology. In actual field study of bird song it is very hard to find a string composing of the MUP that can be identified as a song type beyond individual notes (syllable clusters). For instances the iora produces different songs at different times of the day and in the course of the season. A morning song is composed of two note-types "tweeen, tweun" sequenced either repeating the individual notes, or/and their combination and varying the numbers of repetitions. But each time the notes are produced they display variations in the stress on the initial syllables of one of the notes (tweeen), "weee" or "eee" so that, even if the differences seem to be in the mode of delivery or pronouncing, it is possible to assign two more distinctive notes. By the MUP method each of these strings, either one- noted or two noted will be assigned a song type so that iora can be said to have 2 to 4 song types at least. But it appears that these strings actually make up lines of a fuller verse song and further sequences of these four notes compose stanzas of a same song, *viz.*, all the notes are equally punctuated by a roughly 5 millisecond silence and produced in rougly this 12 to 13 note sequence again and again. The sequences are also punctuated by not more than 5 milliseconds gap. In all, considering the entire season it would be better to say that iora (singing male) has two song types (good song & poor song see section on iora, chapter 2, for spectrograms and explanation) but at least three different songs (morning, noon and after noon) for each song type in his repertoire. It is possible that as the peak season proceeds the songs vary gradually but the song type remains the same until the song type alters at the end of the season so that they have many more songs than three to a song type. My assignation of song types discussed below therefore may not adhere to the method in practice in current ornithological research. But I have derived the labels from Marten, *et al.*, (in Renner & Rappole, 2011).

For instance sunbirds produce song that is made entirely of one or two notes repeated monotonously throughout the day, so do another species of small passerines the flower peckers. This type of singing has been labeled as "endless

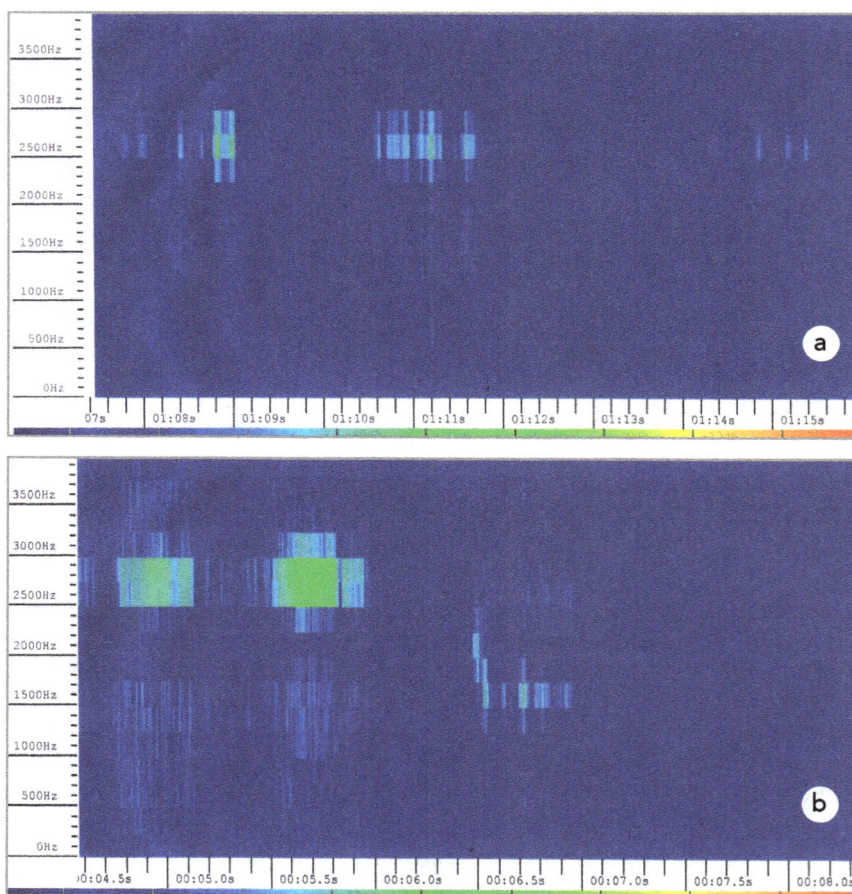

Figures 7a & b: of a note sequence from an iora singing male's early morning and an afternoon good verse song.

singing." Some endless singers such as the female purple sunbird can develop a different song type by changing the notes suitably to their changing breeding status. A second type of song is that produced by birds such as bulbuls and leaf birds wherein the same simple notes are produced with tonal variations. I have called such vocal deliveries as "tonal songs". Tonal songs usually have a more complex notes or syllables than endless singing, but these notes are few and fixed in number. In tonal singing the lack of variety in sound is compensated with diversity in intonal patterns. A third type of song routine is what has been labeled as "verse singing" which are longer songs made up a large number of fixed notes delivered similarly as humans would deliver a song with line breaks or time pauses. A good example is that of the Tickell's flycatcher (see figure below) and the iora (discussed above). A verse singer is also adept at using tonal variation as for instance the Tickell's who delivers its verse song with tonal variations at different times of the day, what I have labeled as "tonal verse songs". A second type of verse song is where a new note is introduced in a new line such as found among the hooded oriole and tree pie. Among both these birds this "true verse song" can stretch across several days even months with each line of the verse delivered across the hours of the day or across several days. Such true verse singing can be found even among the iora and magpie robin (see respective sections for the spectrograms).

All these song routines can be accompanied with movement around the territory, with the song delivered from various locations as well as time of the day. In addition, song types may be sex distributed in a variety of ways. Among many birds the female produces only the one note song such as the hooded oriole. But among some others like sunbirds it is the male that produces the one note song, and the female produces elaborate verse song. There can also be found duet singing (may be considered as a song type) such as among tree pies each sex having its lines in the verse song. It is also possible that the sexes produce songs at different time of the year or day. Among the Tickell's the female indulges in nighttime singing in addition to day time songs. The malakoha pair has an intriguing mode of delivering their duet song: the male initiates the song by a one note "khk" while perched on top of a tree while the female follows with her more elaborate song from under its foliage. In subsequent days both of them develop their songs and deliver by turns.

In addition to having species typical songs, birds in a region sing songs that keep them apart from their own kind in another region. Such variation has been labeled as dialect in the contemporary literature on bird song. A study on leaf warblers in central Asia showed that there are as many songs to a species as regional populations. Songs are distinctive not only to the species but also to a region. Region specific songs are attributed to the capacity of birds to learn their songs by imitation as they grow up in

Picture 10: Dusky Leaf Warbler

a surrounding (Kroodsma, *et al.*, 1999a; 1999b; Nelson 1999; Kroodsma &Miller (Eds.) 1996). They in fact pick up the song from birds and other animals around them which function as social tutor. Even mechanical gadgets such as radio and loud speaker can function as such social tutors. That is why often the bird song in a region will not match with songs of another even if the birds are of the same species. The red whiskered bulbuls are a common bird found all over southern India; but bulbuls in the Bhadra forest have different song pattern from bulbuls in a garden in the coastal town of Mangalore. Owing to the presence of a school near the nesting site, the bulbuls in this region have learnt their song from the voices of little children learning to read and sing every day. Their vocalizations are imitative of the question and answer mode in which children learn their lessons in today's primary schools. Tailor bird is another species given to song learning and produce regional dialect variations by imitation of birds in the habitat such as Tickell's blue flycatcher, magpie robin and even from radio songs. Not only the region specificity, songs can be exclusive to a generation of birds. Younger birds may not sing the same song as their parents because they have learnt their song from different social tutors while growing up.

There are a large number of birds that know their own species specific song as well as produce imitations of songs of other birds, even though many birds may not display any learning at all and sing purely the species typical. As a rule species typical songs are produced early and may not involve learning. But imitations are produced at a later stage of maturity and sometimes even replace the species song completely. Among most species imitation songs are produced by the surrogating lead singer rather than the breeding member. Of course, it is possible that learnt songs can become species specific if it is acquired at the crucial period of speciation event and then by transmission down the generations. To prove this will require study of the genetic make up of a bird and the song itself may be a clue in the history of its evolution. John Alcock in his book Animal Behaviour (Alcock, 2005) claims that there are precisely three bird orders that are song learning. They are psittatiformes (parrots), trochiliformes (hummingbirds) and passerines (oscines) the youngest and latest

Picture 11: Myna (Sarika) is known to speak like humans

evolved among birds (see p. 21). Birds like the cuckoo (cuculiformes) and doves (columbiformes) among the other older bird forms sing equally beautifully out of instinct but the parrots and the passerines like thrush, iora, bulbul, robin, flycatcher, oriole and leaf birds are sweeter for learning their songs.

Studies of bird song have lead to several hypotheses on the meanings transmitted by them. It is possible that birds sing in identity announcement, it is likely that they also sing to communicate with their species and most likely to deter competition over mate and territory. It is thought that song matching is an important means of establishing this sort of communications among the song learners (Armstrong, 1973; Stoddard *et al.*, 1992; Beecher *et al.*, 1996; Mooney *et al.*, 2001). Birds that learn to sing can learn to sing like other birds in its region. Mimicry is the rule of nature. Ornithological research has uncovered three ways in which birds engage in song matching. In song type matching a bird produces a

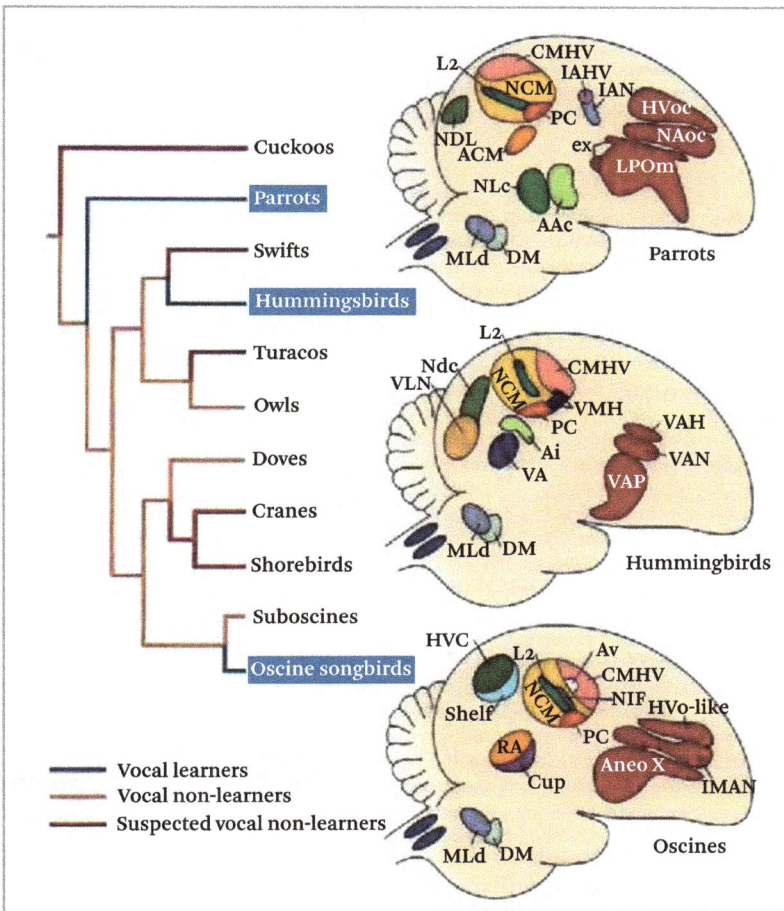

Figure 8: Song control systems of parrots, hummingbirds, and oscine songbirds are distributed throughout the brain in remarkably similar patterns. On the left is the evolutionary relationship among some major groups of birds, including the three orders of social learners. On the right are diagrams of the brains of these groups, with the various equivalent components of the song control system labeled (*e.g.*, HVC, higher vocal centre; NCM, caudomedial neostriatum. (After Jarvis *et al.*, (2000) as cited in Alcock (2005)

similar song type albiet in his own dialect in response to another in the habitat. In repertoire singing the response is in the form of producing another song type from his repertoire. In song mismatching a totally antagonist song can be produced. Accordingly song type matching among birds of a territory is thought to indicate rivalry. Song repertoire matching on the other hand indicates a compromise. Song mismatching indicates complementarities. These conclusions were made on the basis of studies of song sparrows and the sedge warblers (Alcock, 2005 for a discussion). Many good instances are found in the Bhadra region: leaf birds engage in reverse song type matching; tailor bird produces repertoire and mismatching, *etc.* (the data is presented in detail in the sections in chapter Two).

My study of song birds in the following pages provides several new perspectives on the song of birds. The most important of all is the matter of unequal song distribution among members of the same species in a given local population or flock and intergenerational continuity of songs. Not all birds of a flock produce song in a season, neither is song capacity passed evenly down the generations. Fundamentally is the fact that the breeding bird is usually not the singing bird. A second insight is on the inter- species dynamics of songs in a community of birds. This dynamics reveals two kinds of song interphases: a dialectical interphase and a geographical interphase. The former involves systemic shifts in song patterns between two bird species usually genetically related (subspecies or same family) and the latter involves production of pure imitations between species not merely genetically unrelated but even of different class such as animals and gadgets that are native to a habitat. These outlined issues lead me to a different understanding of the song mechanism than in currency at present.

At present it is assumed that breeding bird produces the song and that it is a kind of sexual signaling because his songs indicate his capacity to deter rivals over mate and over territory. This assumption is derived from the observation that birds are bi parenting organisms (or heterozygotic). But if it is true that the singing bird is not the breeding bird like I suggest based on my field observations then the way in which we understand the very mechanism of song should be revised. The singing male must play a key role in a social dynamics whose significance goes beyond a paired reproductive functionality. It is my suggestion that birds sing to hold the flock together to bind them as a species, and least have the effect of finding a suitable mate. Song is a manifestation of herding instinct among birds, primarily for the purpose of reproduction or propagation of species. They give rise to a reproductive praxis beyond pair breeding and possibly an aspect of species coherence.

Further I suggest that the role of song and singing bird has to do with social cohesion at a broader level, *i.e.,* a sense of belongingness to a kinship structure pointing to a coherent inter species dynamics. Songs I think must function at the three levels of a bird's existence, at the individual, his reproductive flock and at the level of the entire community (of birds and perhaps all the creatures) in the habitat. Songs of birds not only give insight into the nature of species (*i.e.,* intra-species) existence but also the nature of inter-species existence. Currently song and song matching are viewed as index to the measure of conflict among birds. I suggest it is important to view the phenomenon of songs outside the framework of conflict and rivalry, whether between conspecific members or among species. Songs should be viewed as means of kinship affiliation and exclusion. Songs organize the social structure among birds on the basis of kinship rather than indicate rivalry.

A key to this perspective is to understand song as negative signaling mechanism; and not as directed by a breeding male at the female but at a younger generation member regardless of sex of the bird. The negative signaling results in the suppression of song of conspecific members, but produces song matching among related species. In both instances song directs the birds towards reproductive cohesion. Field study of bird songs in the Bhadra region furnish us with good examples such as ioras and magpie robins, we find along with the domination of the song of one individual bird - whom I have labeled as the lead singer - occurs the suppression of song of other members in a conspecific flock. Accompanying this unequal song distribution is the unequal distribution of reproductive responsibility so that along with lead singer there emerges a lead breeding bird. But the lead singer has an important role to play because he participates in a breeding coalition by his song. This kind of resolution is accompanied by complete sharing of a territory. His song I suggest initiates the breeding junta and ensures that the reproductive process is completed. It serves to bind and keep the flock together and give it a representation as a unit in the territory.

Pictures 12a & b: A Swallow class lines up on a cable

A question arises if to distinguish between birds like ashy woodswallow, bee eaters, baya weavers, *etc.* which are colony nesters and birds like ioras, magpie robins and tailor bird, so on which form a breeding junta. Ioras for instance are found in two or more breeding pairs. But these pairs can be cross generational while among the former the flock is made up of a "class" of birds probably all precisely of the same age. Perhaps the distribution of song is a clue to understand the difference.

At the level of the community songs enable to organize hierarchy or ordering of the inhabitants. That's why songs tell a great deal about the bird's status in the locale's social hierarchy and his niche (both and basically food and nesting needs) in his habitat. His flock's breeding status can be one of the significant information to impart. A singing bird in a habitat announces the type of his and his species claim over the territory. A territory is important to have for two reasons, nesting and secondarily foraging. I say secondarily foraging because birds especially the small sized singing passerines have a greater freedom to fly about to peck but to nest in a region requires a permit. Turn taking is of course the base rule whether at foraging or nesting. Larger birds will even hustle smaller birds at the foraging sites for better efficacy. For instance parrots force the leaf birds to feed on flowers first so that insects become better visible to their weaker eyesight. Also it is often you will find a discarded nest may be padded and reused. Or a bird more skilled at nest building can be inducted into the community for his or her skills, such as for example the female of the Asian paradise flycatcher is adopted by the orioles.

A song type communicates the nesting habit of the bird. In fact birds are identified by their nesting habits, such as tree hole nester, leaf nester, tree nest builder, ground dweller, mud nester and further tree top dwellers, middle rung, so on. A territory is divided both area wise as well as spatially. Residents generally have complete freedom to use any part of a territory. But this use is governed by the logistic of perpetual movement or traffic of organisms around the territory, usually in a circular shape. Such circular movements are carried on both on every day basis as well as seasonal or annually; so that a location in the territory used by a certain bird in one year is taken over by another species in the next. New comers are inducted into this magic circle only if refereed by a resident or by forcing a resident to give up its place. On the other hand if a resident wishes to back out of his place he may induct another organism in his place (a stand in) at least for the time he is away.

A song bird and his songs play a key role in the intra/interspecies or social dynamics of his entire community. As I said genetic relation or affiliation among them is reflected in the song matching. In support of this in the Bhadra region I found similar kinds of reverse song type matching among sympatric members such as between golden fronted and blue winged leaf birds, between orioles and tree pies. Among three species of bulbuls there is a closer song type matching similar to that between blue winged and NSR leaf birds. Intergrading members are usually repertoire singers who are inducted into the community life to sing for others. The song learning capacity of a fellow bird may be used to transmit to other birds about one's presence and status in the territory. For instance, in the Bhadra region the mobbing behavior of aggressive talkative red whiskered bulbuls is used to fend intruders to the territory commonly by other birds; likewise the alarm calls of a talented tailor bird. While the hiatus of mobbing calls kept up by the white cheeked barbet is in its role of a self appointed territory patrol. Thus potential conflicts and rivalry is re channelized in creative ways to the benefit of the whole population.

In the following pages I have presented close intense field observations on the orange headed thrush (*Zoothera citrina*), common iora (*Aegithina tiphia*), oriental magpie robin (*Copsychus saularis sualaris*), Tickell's blue flycatcher (*Muscicapinea tickelliea*), four species of leaf birds (*Chloropsis cochinchinensis*, *C. aurifrons frontalis*, *C. aurifrons insularis* and *C. aurifrons frontalis* NSR), Racket tailed drongo (*Dicrurus paradiseus*); hooded oriole (*Oriolus xanthornus*), Indian tree pie (*Dendrocitta vagabunda pallida* and *parvula*, Jungle, and scimitar babblers (*Turdoides striata malabarica* & *Pomatorhinus horsefieldi*), red whiskered bulbul

(*Pycnonotus jocosus*, red vented bulbul *Pycnonotus cafer*), white browed bulbul (*Pycnonotus luteolus*); purple sunbird (*Nectarinia asiatica*) and purple rumped sunbird (*Nectarinia zeylonica*) and grey breasted prinia (*Prinia hodgsoni*); Asian koel (*Eudynamys scolopacea*), Indian cuckoo (*Cuculus micropterus*) and Eurasian cuckoo (*Cuculus canorus*), Blue faced Malakoha (*Rhopodytes viridirostris*), white cheeked barbet (*Megalaima viridis*) and coppersmith barbet (*Megalaima haemacephala*), and spotted dove (*Streptophelia chinensis*). All these bird species are widely distributed through out the Indian peninsula. But the data presented herein is specific to those subspecies whose range is limited to the southern peninsula (see appendix One Table A2). Further the actual field observation is limited to the Bhadra forest (sanctuary area of 492 sq km) and its outskirts. The forest situated at Lakkavalli makes up a small part of the Bhadra river basin or catchment in Chikamagalur district in Karnataka.

All the birds listed above are found in a territory of 230 sq km that makes the epicenter of the total area under observation for this book. The total area inhabits around 205 diverse species (see Appendix One Table A1 and Table A2). The epicenter is made of a diversity of ground types being a hilly area with cleared forest but skirted by deeper open jungle (*i.e.,* the Bhadra sanctuary forest) on one side. There is intense human activity in much of this territory. The observations were made over four seasons starting February 2011 to September 2014. Among these birds only the thrushes are non residents and are migrating. They visit the territory only in a cycle of two or more years. Many of the birds under observation belong to different territorial areas (in the several directions) but that which overlap with the 230sq km one, *viz.,* parvula tree pie(north east), insualris and NSR leafbirds(southwest), red vented bulbul (northwest). But parts of the territory under observationare used for breeding in some seasons or only for foraging (NSR leaf bird example).

2

Singing Passerines

Orange Headed Thrush

In classical ornithology thrushes are much celeberated as song birds among passerines. These birds of a large family clade are widely distributed globally, are seasonal and long distance migrating. Among the birds of southern India studied in this book the thrushes are one of the truly migrating species. Ali in his The Book of Indian Birds (2002) records that they migrate to and from the Himalayas, flying as far as the Western and Eastern Ghats in winter, even as far as Sri Lanka. In comparison when we are talking of other locally migrating birds such as oriental white eye, bush chat, or crimson sun bird in the region we are merely referring to short local migration from one territory to another in the same region, some 20 to 40 kilometers across not more than that. Thus relatively the thrush in this region are long distance migrating even though a far cry from the inter-continential migrating birds such as Siberian cranes. The thrush arrive in the region in early spring for the purpose of breeding. There are five species of these birds coming here cyclically depending on the span of migration and all of them produce melodious song. But the songs are heard only for a brief part of their breeding season. The songs are produced well after the breeding is underway, rather than at the start of the reproductive season, like so many other species. My field observation indicates with certainty that all the five species breed here even though it is puzzling why Ali (2002) records them as merely winter vistors to the region. I think their pattern of migration has lead to this erroneous observation as we shall see below.

Pictures 1a & b: Blue Rock Thrush & Blue Headed Rock Thrush (male)

Since the thrush are migrating birds their habits have drawn much research interest. Charles Darwin in his book The Descent of Man holds their habits as proof of his theory of sexual selection. He concludes that the songs of the thrush must be evidence for them as male adaptations acquired to win over a female. As proof he claims that the male birds migrate first and reach the destination first so that he can prepare the site for reproduction. But my field observation indicates the contrary. Their method of migrating is crucial to understanding their nesting habit. There is no doubt that they are migrating species but they do not fly the distance at one go. A single migration can take several years, with the flock taking sojourns at different nearing sites on the route. That's why in a flock there are always new fledglings that are native to the region for having bred here. These new fledglings have to wait till they grow into adults before they can join the migrating flock, which in the meantime will have arrived in the region. It appears that a newly fledged young adult female thrush does not join in the migration until she has completed at least one reproductive cycle (that is she migrates only with her first chick(s)). It is the male that migrates to a nesting site to the waiting female but likely only from a nearby territory. In addition it appears that thrush are tree hole nesters and there is not much nest building. Babbler species such as streaky breasted, tawny bellied, yellow fronted and rufous fronted babblers also display a similar migrating pattern, but they do not fly the distance. Since they are basically ground birds flying in short sprints they migrate from one forest floor to another. It is possible that the orange headed thrush can take recourse to such a method if the need arises.

In the following I shall present the field observation I made on the orange headed thrush (*Zoothera citrine cyanotus* species that was sighted). These birds arrive here in a large flock of about twenty birds in April and sojourn here up till September. In this last month one is likely to find a lone orange headed thrush lingering on after the flock has taken flight. Since they are in a place for at least six months of the year it is likely they cannot visit the same place annually. This explains why in the years I kept a watch they were here in April 2011 to September 2011. But they did not return in the next year April 2012; niether did I sight them in 2013. But they were here in April 2014.

Among the orange headed thrush there is only one pair that sings in the flock of about twenty birds that arrived here. Both the birds in the pair produce song but on different occasions and never together. These songs are delivered from middle rung of a tree. The rest of the flock remains stone silent through out the season. These birds are not strongly sex dimorphed. According to the literature the female is duller blue and the white patch on the face is smaller (Grimmett & Inkskipp, 2009, vol. 9, Ali & Ripley, 2001). She is also slighter in size the male is more rounded. I have assigned sex identity to the two birds I saw singing based on this

Picture 2: Orange Headed Thrush

reference. Regarding the breeding status of the singing pair, it is likely that they reproduce only the flock's lead singers each in every alternating season while the several breeding pairs (approximately 10 pairs) that make up the migrating flock reproduced the flock size. There is a parallel occurrence among the grey breasted prinias that engage in a similar reproductive economy (see last section in this chapter). As I noted above the other members in the flock did not produce any vocalization.

What is intriguing is that there appears to be sex role polarization in the meaning and purpose of the songs that are sung. Even though both the singing birds produced the same song that is highly musical (tuneful) and an elaborate composition made of several notes on the two different occasions; it is possible that the intonation pattern on the end notes was different "twetwitutwitwuwitawu?" sung by the first bird and a more level pattern sung by the second bird. But this cannot be accounted as differences in song capacity. Most significantly the male thrush seemed to sing the identity song upon arrival while the female does what I have labeled as "decoy singing" as the birds dispersed in the territory having settled for a nesting site in a locale in the territory.

The first bird that sung I presume was male and the song was delivered when the flock actually entered the territory and surveyed it from one end to the other. This song seems to fit the identity announcement theory about songs of birds (Alcock, 2005). (In fact I was alerted of their arrival by hearing this intriguing sweet song and rushed out to find out what bird it was.) But the second song produced by I presume the female bird when the birds were flying in the opposite direction on the next day during the same hour distinctively communicated a totally different meaning. This second song resembles that of the singing sky lark recorded in ornithological literature. Not only is it found that sky larks sing in the sky when flying they do it when threatened by a possible aggressor. The female thrush displays a similar kind of song decoy singing as I call it, used to distract the observer away from the flying flock and give the accompanying birds time to fly while the observer's attention is so diverted towards the singing bird. As I stopped to stare at these prolific birds, the female of the lead pair separated from the flock and fell behind to sing loudly straightly at me and I stood there transfixed by the sweet melody. Before I realized the rest of the flock had flown away and I forgot to see in which direction.

This kind of role distribution of songs among birds of a breeding flock, turn taking engaged in singing and the differential song types suggests that we should revise our generalizations about songs of birds. Songs seems to be more a part of a social or group dynamics.

Iora

The iora is a small sized (about 11cm) passerine having an expansive range across the entire Indo Malayan eco zone (even though never found outside this range) and may be sighted almost all over India. These birds usually move about in small parties of two or three breeding pairs. Not all the members of such a flock can sing. There is a lead male singer but he is not the breeding bird, The rest who are the breeding members can only produce at best a poor imitation of his song and that usually only at the start of the breeding season.

The iora singing male is the master performer of any South Indian orchard where his songs can be heard all the year through. He produces two versions of his song in the same season, one version is the nostalgic elaborate verse composition of the peak breeding months (assigned as good song type and species typical) and a second version is a defiant challenged sounding poorer imitation of the first produced at the end of a season (poor song type and atypical). He continues to sing the poorer second version for the rest of the year (spectrograms are given on next page). Song performances are delivered from various parts of the territory from under foliage of medium shrubs starting as early as six o'clock in the morning the last song of the day comng at high noon around 4-30 to 5 o'clock. He has several different songs in each of these two song types, such that he produces distinctive songs suited to the time of day, dawn, mid-morning or noon and late afternoon or evening (see discussion in Chapter One and spectrograms in Figure 2 there). The song types remain invariant across distinctive populations of ioras in different territories and shows evidence of being species tyical. In addition they also remain invariant within a population intergenerationally, with perhaps minor differences produced in generational cycles.

Not only is there only one lead singer in a community of ioras in a territory. In addition, the intergenerational continuity of song capacity and breeding capacity is broken with song unequally distributed between birds of two different seasons. Only a bird from every alternate generation turns out to be the perfect singer who can sing the masterful typical song of the iora species. The genetic relation between such two singing males if we were to put it in human terms is comparable to uncle-nephew. This intergenerational continuity and the iora lead singer's inability to keep up the best version of his song and its subsequent variation seem to give a clue to understanding the song mechanism among birds as negative signaling. It is my suggestion that at the end of the breeding season when the fledgling is out there takes place suppression of the song mechanism among the iora flock, in the singing bird but more importantly in the new fledging member. And that this sets in motion the reproductive activity for the next season and secures species propagation.

The ioras are communal breeders of a unique kind and show strong cooperative intra species dynamics.They are found in small parties of two or more breeding pairs. Nests are built in large ever green trees like jackfruit; or in the undergrowth of

Picture 3: Iora lead singer in August (into his second season)

bushes such as bougainvillea or lantana. Foraging is also done in small parties on tall trees as well as in the undergrowth. The lead singer participates minimally in all these activities and prefers to sing in solitude from medium sized bushes most of the year. His female remains totally integrated with the flock throughout. Peak breeding season falls from February to July every year.There is much written about the courtship displays of the male iora but it is very likely that one will find the female executing these displays in the beginning of their breeding season. She hangs from branches upside down and twists and turns or puffs out her feathers in a male like posture. She is second only to the leaf bird female in display of such abandon. But in the contest of songs the male that makes breeding the second priority sings the best song, she and older males in the flock produce a feeble song which soon becomes extinct as the breeding season gets underway.

The ioras are non migrating and reside in strongly demarked and quite an expansive territory within their habitat, edges of forest clearing and light forests with evergreen trees like sandal, jack, jasmine, and honge in the Bhadra region. Within their territory, they can have different breeding and feeding areas and show free seasonal movement. These movements traverse a circle around the territory in the course of each annual season starting in January-February. This movement and the intergenerational variation in song, molting patterns and possibly a mid season clutch in some seasons (see discussion on next pages) in the course of the year also lead to much confusion about their diversity.

The literature claims at least two distinctive species named common iora and Marshall's iora (*Aegithina tiphia* and *Aegithina nigrolutea*). Salim Ali in his ten-volume Handbok of the Birds of India and Pakistan (vol6, Ali & Ripley,2001) lists five subspecies under common iora identified by the degrees of colour differences in the males' breeding plumage, more black, green or yellow on upper side. Actually distinguishing between the subspecies by coloration is tricky. According to his book the diegnani subspecies breeds in the southern peninsular region. But subsequent literature (Grimmett & Inkskipp, 2007) identify only humei and multicolor in

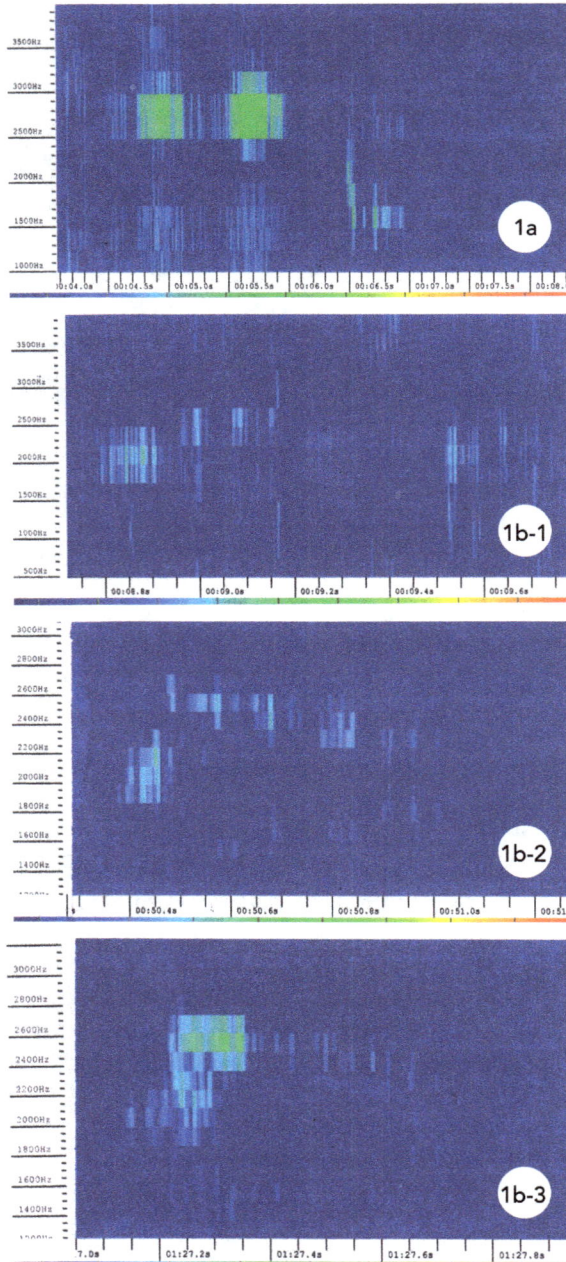

Figures 1a & b: A comparison of the two song types- *i.e.*, good & poor- of the iora singing male. Spectrogram in Fig. 1a shows the fully formed notes from a late afternoon peak breeding good song ("tweeeee, tweun"). Figs. 1b-1, 1b-2, 1b-3 show the notes ("twei...twee" "twei...wei" & twee...wei") from an early morning poor song during post breeding. Note the two song types are sung at different frequencies, and cannot be explained by changes in allopatric conditions the region. Considering the ambiguity in nomenclature, and by my own field observation it seems there are only two species of ioras and those identified as subspecies are not actually so. Not only does the molting pattern as well the ability to sing vary between two generations of birds it also varies in the course of the year in the same bird. In the Bhadra area in the four seasons Feb 2011 to September 2014 I kept a watch I conclude that the ioras I observed all belonged to Common iora but they were in different seasons. And that a community of ioras in a territory can have such cross generational pairs that have different breeding and singing capacities. I did not find any member of the Marshall's iora (*Aegithina nigrolutea*) species.

The song routines of the ioras reveal a most fascinating intra species dynamics in the iora community. In the first season there were two pairs in the party/flock in the second season there were at least three pairs of them (one new younger pair and two from last year's. But out of the latter the older pair from the previous year became invisible after season initial foraging and subsequently there were only two pairs visible.) It is curious that I did not observe any contest over mates all the birds were already paired off. The ioras start to molt as soon as they enter breeding in January-February. This leads them to lose their greenish yellow plumage colouration and to temporarily become completely black with only the white wing bars on the upper side. Now it may be noted that they do not become fully fledged singers until the male attains his amount of black on cap. In the breeding party the bird that molts first gets to sing. On the other side the breeding bird attains the maximum amount of molting. In the singer bird the molting is arrested and becomes limited to the cap only.

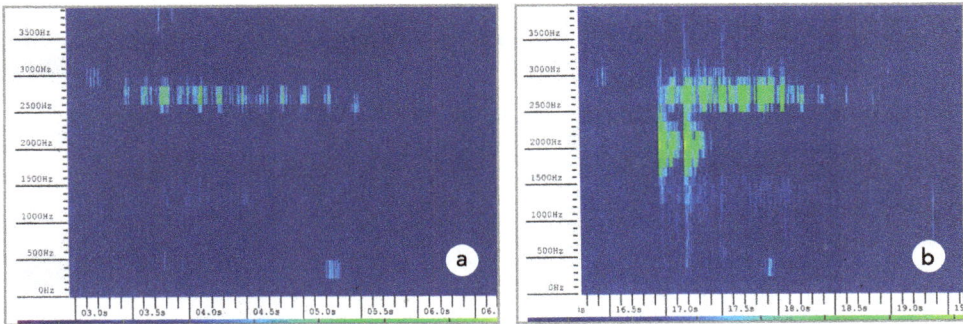

Figures 2a & b: Two lines from the imitative but distinctive song of the iora breeding male, ("twitwitwitwitwittwitwitwit" & "tweeun..twitwitwitwitwitwit"). This song is produced without variation at different times of the day

In the first season the younger bird entered molt sooner. After this he broke away from the party and found a perch from where he can sing solo undeterred. His female moved about in the company of the other pair(s). He can sustain this good song type for as long as from February up to July after which the notes deteriorated to the poor song type even though he continued to defiantly keep up the singing till the beginning of the next season. The deterioration in song is accompanied by gradual reversal in molting. By August the lead singer has only an eye stripe (see picture above). This enables him to pose as a new bird. The actual breeding older male did not produce any song.

In the second season even when there is a younger pair the last year's singer continued to hold the stage. In fact he still started the year sooner than the younger birds by achieving black on cap first. In comparison to the intensity of songs sung in the first season in the second season the lead singer spaced his songs far apart interspersing them with his poor imitation. That is to say in the first season his songs were more frequent while in the second season fewer in number but kept up for a longer period so that he delayed the loss of song capacity. In addition the molting on cap was arrested sooner than in the first season.

A strangest phenomenon is that in this second season instead of a contest of songs so much celeberated in classical Ornithological literature the younger bird of the new season which molted strongly seemed to shun the older singing bird. He hustled all the females in the flock (even the last year's lead singer's). He produced a poor song type distinctive from the lead singer's poor song type (with two notes 'tweeun' and 'twit' produced in fixed numbers of repetitions; see Fig 2 above for the spectrogram) that seemed to be a mockery - technically song mismatch - of the the lead singer's of the lead singer's both song types. This song performance was also delivered with a subversive ritual from top of tall trees rather than hid under bowers like the singing bird does. By the by he molted completely and began to look quite unlike himself a black and yellow bird. He now gracefully gave up his imitative song and settled to his breeding role while the lead singer continued his verse song.

Pictures 4a, b, c, d, e & f: Stages of molting in Iora breeding & non breeding females(a & b) and in Iora breeding (c, d, e) & singing (f) males

Some interesting facts need to be emphasized here which lead me to the conclusion that the lead singer is not the breeding male among the ioras of a territory. All the ioras were strictly paired off from the beginning. The ioras are not sex dimorphic and look identical at the beginning-end of each season. But as the season proceeds the females are distinguished by the complete absence of molting in younger or minimal molting in older females. This identification is corroborated in the literature on ioras. Three kinds of foraging were observed: individual, paired and communal in addition to other activities like collecting of nesting materials. Individual foraging was found only among the lead singer and his female. The lead singer was found never to participate in the paired activities like collecting of nesting materials or in-breeding foraging. Neither did he participate in the initial community foraging for long that took place in both the seasons which was lead by the breeding bird. His female was not found collecting nesting materials, which seems to be mostly a pair activity but she joined in the foraging.She contributed a great deal to the feeding of th female chick (the next generation breeding female) born the mid season. Most importantly the period of intense singing of the lead singer coincided with the peak breeding season, late Feb -March when the egg is laid and June by which time the chicks are hatched in both the seasons. All this time he stuck to his solo performance always, in several locations within the territory.

It is clear among the ioras - and perhaps in most other birds - there seems to be a choice between biological investment in singing or reproduction. A bird that opts to sing falls behind in reproducing. Of course this does not mean that the role of the song is only marginal to the reproductive praxis among the ioras. It is important to see that birds in a flock participate in a breeding junta. The contest seems to be between song and breeding roles rather than over song between the two birds. As for the singer he has relinquished his reproductive role. But his contribution is to enhance his flocks' reproductive success by herding them together and keeping them together as flock. Perhaps this also enables them a claim over the territory. His song seems to function as a psychological device of group cohesion and his role being of a charismatic leader. This distribution of capabilities and roles point to an economy, which is far more efficient than if both singing and breeding were done by the same male. It makes sense for little birds to engage in such a breeding junta.

Among the ioras song distribution seems to be guided by the intergenerational gap with a bird of every alternative season producing the melodious typical song of the species. There is a possible explanation for this. In current ornithology (see Alcock, 2005) song learning among new fledged members is attributed to the presence of a social tutor, even though the species typical song is thought to be transmitted genetically. In this view the lead singer must play the role of social tutor, even though this role cannot be described in a straightforward way in terms of song learning. His intermitting capacity for song combined with reproductive inviability must prove as negative signaling to a new member so that his libido is turned away from song and towards viable reproduction. But why is it that the song capacity is handed down in every alternating generation? It may be noted that in the season the lead singer produced intense species typical song abruptly losing it at fledging time the new member did not produce the species typical song. But in the season the lead singer produced less intense species typical song frequently and throughout interspersed with poorer imitation produced the next generation lead singer. Thus the alternating intergenerational continuity of song can be correlated to the differences in the capacity to sing of the lead singer in each season. This also corroborates the claim that species typical song is not learnt but the new member merely learns to recognize what he already has genetically in his possession. The role of the lead singer who can produce the species typical song is more symbolic than real. In other words song mechanism among birds functions by song suppression among fellow members directing them towards reproduction away from song.. Among the ioras this occurs every alternative generation, coinciding with the extent of the lead singer's capacity for species typical song.

There could be a second possible explanation for this kind of alternating continuity in song capacity. I have said that the iora breeding season falls from Feb up to July every year. This is not an accurate observation. An adult iora breeding pair has two successful breeding seasons both of which start in February in two consecutive years. But in between these consecutive clutches they also produce a mid-season clutch in October-November.Two eggs are laid in the first season; the mid season and the second season produce a single egg clutch each. I suggest that this is the minimal reproductive out put for a breeding flock, which seems to be the bottom line for the perpetuation of the species. The first season invariably produces the singing male and the breeding pair comes from the mid and second seasons. The lag in time between the laying of the eggs and fledging of the chicks by the distribution of offsprings over three seasons seems to have an impact on the song capacity of the males. In the first season the eggs are laid at an interval of about one month each. The egg that is laid first and fledges first turns out to be male. The second egg turns out to be female. Subsequently the male fledgling reaches complete song capacity long before the second chick (his female) can fledge. And in the interim he turns to singing. He continues to sing even after the female reaches her adulthood but they do not reproduce viably. But ideally for him the parent birds

are still in their prime and go into breeding once again by now and produce their mid season second clutch. In this second clutch (very likely a single egg clutch or else there is a loss of an egg) the breeding female is produced (this seems to be fixed with a certainty as I observed over three seasons). For the breeding male to be born they have to wait till the next season February. So the breedeing male is always younger than his female. This pair got out of two separate clutches turns out to

Picture 5: Iora breeding pair at foraging and collecting of nesting material

be the flock's breeders and the male produces only a poor imitation of the song and only for a short period season initially. It appears that the intense song of the singing male (first born) in his first season has the effect of suppressing her song (and his own female's) and ensures that she turns to reproduction. Likewise the subsequently born second male chick's partial capacity to sing for his situation can be seen as caught between a female that has reached breeding capacity and an older singing male who is reproductively inviable.

There could also be a possible explanation for the variation in song capacity in the same season and sexual inviability of the singing male iora. Among most birds there is a period between fledging up to attaining sexual maturity called as juvenile stage. A juvenile bird cannot reproduce since he is immature. Among the ioras the singing male reaches full song capacity in the juvenile stage. Subsequently he and his female fledged a little after him enter breeding in the off season in October. But since both of them are still in the juvenile stage they do not produce any viable egg. After this up to February March the start of the actual breeding season the singing male bird keeps switching back and forth between his good and bad song. But by March his molting is arrested with only a tranquil black cap and he achieves perfect song once more (molting generally starts with the nape turing a livid red). He has also reached complete adulthood but remains reproductively inviable.

Perhaps it is worthwhile to compare the singing bird in his incapacity to reproduce to the neuter or (eclipsed male) referred to in much ornithological literature. But his incapacity is not due to any sexual contest between a stronger breeding male but due to his song capacity, He seems to be forced to make a choice between singing and breeding. Yet he seems to play a far more significant role in the propagation of species. I suggest that the role of the singing male points to the social dynamics in bird communities that is comparable to W. Hamilton's theory of Kin selection (1964). Turn taking is of course a base line criterion for sharing the resources of a habitat. But the ioras clearly demonstrate that the dynamics is more determined by non functional criteria such as charisma, leadership imitation and altruism. The altruism of the lead singer brings about species cohesion beyond pair breeding and species propagation. The case of the magpie robin discussed below parallels that of the iora and corroborates the prediction on choice between breeding and singing but of course with an interesting variation.

Oriental Magpie Robin

Traditionally the robins are considered ace singers among passerines. Almost every other field observation records the breeding male song. In the Bhadra region there are sighted two species of these birds. The oriental magpie and Indian robin both are local residents here and prefer same kind of ground type covered with evergreen trees which border open land. The two species steer clear of each others' territory and occupy separately permanently and exclusively round the year their chosen area. This territory is wide enough to provide foraging area, nesting site as well separate roosting site. Robins are not much of nest builders. The Indian robin is found to build rough cup shaped nest with twigs. These nests are found in the middle rungs of tall trees with leafy canopy close to rich forging areas in a secluded spot within the chosen territory. The magpie robins on the other hand live in nests made of twigs or grasses inside tree hollows. The tree hollows may be some 15 feet above the ground.

The Indian robin (*Saxicoloides fulicata*) and oriental magpie (*Copsychus saularis*) reveal distinctive capacity to sing. The songs of these birds, like that of the iora discussed earlier, support only partially the hypothesis that bird song has a significant role to play in mate choice. My field observations reveal that a robin female in both the species sighted in the Bhadra region has obvious use for her mate's song and seeks him to play the typical male protective role. Surely it should follow that she will invest in a mate who can sing charmingly and thus ward off predators. But on the other side it seems that the singing bird among them is a non breeding and a younger male. This younger singing male is inducted into a polyandrous breeding trio in which he displays what I have termed as 'decoy singing' as well as territory songs. He seems to be inducted into this polyandrous set up only for his singing capacity and not for his breeding prowess. But he seems to play surrogate-guardian role as the chick(s) are being hatched and fledged.

A most remarkable feature of the song of the Indian robin is that the female uses her mate's song as cover from intruders into her territory. Among them I recorded only the male's song. The female appears not to have this capacity. The Indian robins are sexually dimorphic. It is possible the female uses her distinctive cryptic plumage, subtle dull shades of blue and buff, red vent and eye ring as camouflage for species identity. She is generally found hustling the younger singing male to face intruders and to distract them by his song. On my field watch I was met with a small family drama with the domineering Indian robin female goading her reluctant male smaller in size to sing. In addition I did not observe any interaction with other species in the habitat. The song also appears to be typical rather than learnt.

The case of the oriental magpie robin is even more intriguing and in the following I review my observations in much more elaboration. Among them, birds from two consecutive generations reveal different song capacities in addition to variations from one season to the next in the same bird. In a season from March to July several different generations of magpie can be heard singing their different songs in the same territory. The breeding male and the female in their first season and the surrogating male (in his first season or second) all of them produce their songs as if in an orchestra. Perhaps it is owing to this, in classical ornithology it is theorized that there is a contest of song among the breeding birds in a territory and that this contest is crucial to the female's mate choice. To the contrary my field observations reveal that the magpie robins participate in a polyandrous breeding junta which includes a surrogating male and a breeding male both attached to the same female. In addition at any given season there are always several of these breeding juntas in their different seasons in the same territory. These birds make up the local population of interbreeding magpie robins. Another interesting factor is that there is no sexual polarization in song capacity among the breeding male and female magpie. There is in every alternative season a female in her first season who is also a singer in this junta who can sing the species typical song even better than her breeding male and like him only in her first season. Like the Tickell's blue flycatcher female (see next section) she also contributes to nighttime territory monitoring during hatching time. The magpie robins do not produce region specific dialects and their songs are uncannily identical in all regions, example in an orchard in the coastal town of Mangalore and in a forest clearing of the Bhadra region. But it is possible that the breeding seasons of the adjacent populations varies along a clinal gradient, with members in Mangalore starting a month later than the Bhadra, May to August as opposed to March to July, respectively

An adult oriental magpie robin resident can be identified in the garden habitat by its sharp one note "tweop" produced repeatedly

Picture 6: Magpie fledgling

whenever there is an intrusion from other foraging birds in its territory. This one note call is produced by adult members who have completed their breeding season. But a younger bird in his/her first season at the start of first breeding season in March-April produces a fully fledged song "twe tavi twe tavi"sung for about four weeks any time before fore noon of a day. Typically the male delivers his song from the top of young trees about 8 feet from the ground and if it is a female bird then perched on jutting branches at the middle rungs of trees.As the season proceeds and nesting gets underway the singing (who is also the surrogating) male takes over. In him the song develops into an elaborate passionate rendering kept up for long stretches, approximately half an hour at a time sung from amidst the canopy of very tall tree tops. This singing male's song is composed of several notes "twe tavi twee wee ee"sung in a fashion the notes broken up into their components and repeated in various combinations what I have called as repertoire singing adapting a nomenclature from Alcock (Alcock, 2005, p. 29). This song is delivered in the manner of a vocalist in a street show or procession, with whistles and toots thrown in and delivered at a fast pace and ending in a crescendo. The magpie seems to play the role of a ring master in a monkey show.

This song may be sung several times in a day from mid morning to high noon. During this time he flies from location to location within his territory addressing his songs to specific birds near their nesting sites, such as cuckoo, leaf bird, Tickell's blue flycatcher and flower pecker. At the end of these rounds he returns to his nest holes - he has several scattered across the territory- and sings there also. Near his nesting holes the song is simpler without the whistles and performance. On this occasion the female may be present near nest holes as silent witness. An inspection of a nesting hole frequently visited by the singer and his female revealed that it did not have any eggs even during peak singing (peak breeding season). That's why I decided that the singer was indulging in decoy singing and mock nesting routine that I observed even among other birds (discussed in the section on leaf birds)

The tree top singing of the surrogating male in his first season comes to an end by July. He will not sing any more in this season. But in this season in his absence the breeding male replaces it with a one note "twee." This one note is delivered from the lower rungs of trees some four feet from the ground. The surrogating singing male will resume his song only in the next season, in March again. But then instead of the tree top singing he produces another type of song, the same notes of the last year sung differently and delivered from the middle rung of trees. This song is kept up through out the season, sung intermittently during the day. The breeding male has stopped producing any song by now.

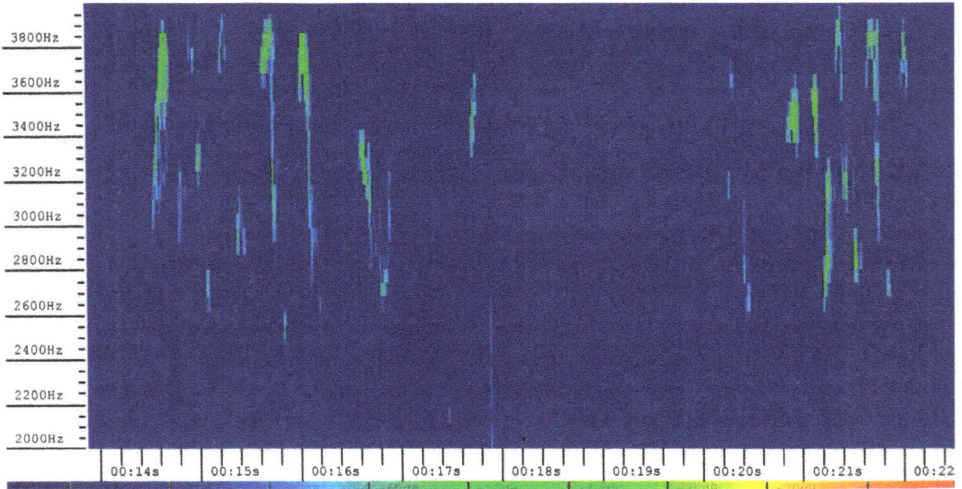

Figure 3 : Spectrogram of a 2-stanzaid song from the repertoire of the surrogating robin during his second season peak breeding/ in-fledging (spectrogram analysis of first season peak song is not available). The member produced about 25 such songs in 5 minutes that appear to be a part of a complete narrative sequence that ran up to 15 to 30 minutes. Each song has two stanzas varying from one line to five and seven lines. The first stanza made up the body of the song narrative with imitative references to one or many other bird species in progressive order in the territory followed by a second concluding stanza sometimes with a one- line conclusion but always with a pattern of notes comparable to a word list in children's school language book. It may be conjectured that the member picked up his song routine from a school in his territory

The surrogating magpie male can be said to produce three different kinds of singing spread over two seasons, one decoy singing near his tree nest hole, peak breeding repertoire singing from tree tops and a post hatching or in- fledging song. In comparison to him the breeding male produces far less quantity, duration as well as song quality.The decoy singing of the surrogating bird is of a simpler nature and closely follows the species typical song also sung by the breeding male in his first season. The in-fledging song (in second season of the bird) is an elaborate verse song carried on through the day over the entire season.The repertoire singing (first season peak breeding) may be said to be an elaborate song kept up for half an hour at a time where the same syllables of the species song are reorganized and sung to a different rhythm, pitch and tone. A similar kind of repertoire singing was observed among the tailor bird and parvula tree pie and have been compared in Table Five in the section on tailor bird below.

The unusual song distribution between members of a same species that occupy the same territory is of significance to our understanding of the song mechanisms of birds. Among the magpie robins, the most impacting factor seems to be the female's choice of more than one mate in the same season akin to polyandry. Here again we find a correlation between extent of reproductive participation and singing capacity of a bird. As I said above singing capacity differs between two consecutive generations of magpies and also from season to season in the same bird. Unlike among ioras every male magpie robin can sing and produces song in his first season. But song capacity varies between birds of two different seasons correlating to their participation in the breeding junta. Therefore there is only a partial intergenerational continuity in song.

The breeding season of the magpie in the Bhadra region falls from March to July. The breeding male and female produce a brief quantity of song in their unpaired state before they enter mating. The peak singing of the surrogating male magpie coincides with the peak breeding season of the breeding pair. In the first season he produces intense song for the peak breeding period; in the second season he produces extensive song spread across the whole season but sung intermittently. I suggest both of this is suited to his reproductive contribution, which in the first season is fledging of the chick rather than mating and hatching and perhaps extends even beyond this because the young juveniles are under his tutelage, especially the next generation surrogating and female birds until they reach breeding age.In the second season it is limited to incubating stage and the breeding male seems to have a greater role in the post hatching during this season.

Body posture plays an important role in social communication of birds. Perhaps one can hypothesize the universality of certain postures typical to the many birds (a kind of bird body language) such as threat posture neck feathers puffed out and head jerking; or posture of prideful paternity/maternity beak pointed upwards neck straight; or posture of femaleness body balanced like a boat on its two legs, face exposed up to belly in front; posture of surrender or abandon wing loosely hanging on either sides; posture signifying need for privacy cleaning neck feathers with beak; posture turning tails showing the vent by lifting the tail upwards; whirring of wings signifying familiarity (see picture grid2 Poses of Birds in chapter One above). This is of course not an exhaustive list; more observations need to be carried on this front. While singing body posture may also contribute to the total semiotics of song. The seasonal variations in the magpie robin's songs are also accompanied by variations in body postures. The off season (tweop) and season end or infledging songs (weee;tweee; twee tavi ee ee, *etc.*) are carried on with the typical body positions of robins: tail pointing upwards wing spread out a bit on either sides, this is the typical style of robins as a species. The tree top singing of the peak season is done with the tail placed straight downwards and wings are pressed tightly to the sides. It would be of much psycho biological interest to study the spatially differentiated location on trees the magpies chose to deliver their different songs.

The songs of the robin seem to upset the popular hypothesis that breeding males are the real performers. Just as among ioras it seems more that they take part in a breeding junta or coalition which is mutually benefiting and fulfilling. The surrogating singing magpie robin seems to have an extra large role in the breeding coalition he sings as well as surrogates. Two questions rise in this regard: Why should the magpies require a surrogating male for successful fledging of chicks? What is the meaning of the repertoire singing? A magpie breeding pair is formed out of members from two different populations in the territory. The breeding male comes from a different parent clutch and the surrogating male is from the same clutch as the female. It is observed that such fellow members display notable morphological variations such as differences in beak shape and wing and tail size. I also observed that the two clutches were not raised in the same location within the wide territory. Invariably each new generation is fledged in a different location, utilizing maximally the variety of niches available such as hilly forest clearing, edges of plains or grassy scrub, and edges of lakes, perennial water bodies, so on. It is possible that this magpie robin interbreeding populations form a cline akin to ring species (Renner & Rappole, 2011, p. 90). It appears that micro differences in ground type also give rise to differences in the quality of nutrients available, climatic conditions *etc.* and enable different adaptations. This is strongly supported by the differences in body size & length of tail of the breeding male and the surrogating male and also in beak size and shape in every season.

Three subspecies are listed for magpie robin in Ali's survey (vol 8, Ali & Ripley, 2001), *viz.,* Copssychus saularis saularis, C. s. ceylonensis and C. s andamanensis. Out of these the first two subspecies are listed for Southern India. Ali distinguishes these varieties by subtle differences in the tint (greenish versus bluish sheen) of the otherwise typical black and white plumage and the extent of black patch on the

Picture 7: Different beak types among Magpie robins in a single population: Pictures 7a-1 to 4, 7b-1 to 5 & 7c- 1 to 3 show beak differences among females, surrogating & breeding male magpie robins across four consecutive seasons

chest. The members found in the territory in my field observation do not display the difference in the tint (or sheen) of the black plumage nor any remarkabe differences in the extent of black on the chest. So it is unlikely that these morphological differences are subspecies (used strictly to refer to non interbreeding flocks) differences. I suggest that it is more likely that the local population of the species maintain a certain complimentary adaptations and genetic diversity. For instance at least six different beak shapes could be identified among the same interbreeding magpie population: long pointed tapering beak with hooked tip, blunt pointed short beak, wedge shaped beak with inner edges of the lip curved like flycatchers with hook and without hook at the tip, wedge shaped long pointed beak, wedge shaped thick beak with the blunt tip curved inward, In addition these beak shapes vary in size and length between male and female, and surrogating and breeding male. These morphological features are distributed among the members of the breeding junta so that at least two to three types of beaks are available to them simultaneously to each junta.

Magpie robins are specialists that primarily feed on termites. Termites are found in two forms with wings in the wet months and without them in winter. At the outset it would appear that the beak shapes must enable the magpie to maximize use of this specific niche. For instance, the long pointed tapering beak with hook and the wedge long pointed beak is used to pull out the termites that infect forest trees. Wedge shaped beak with hook may be used to pinion breeding winged termites found on moist forest floor. Yet as field observation indicates that this cannot explain the diversity fully because all the beak shapes are of equal efficiency in locating and consuming the insect. It appears more likely that the differential beak structure is maintined for reproductive feeding and is group selected. When in a season both the breeding male and female has wedged shaped beaks (with or without hook) this can be disadvantageous feeding the chick. Not only that unless a member is inducted with blunt thick beak (or tapering) to retrieve foodstuff stocked in the gullet pouch of the parent, members in subsequent generations will develop more and more wide beaks making it even more difficult to feed. So magpie robin members that belong to the same stock and that belong to the same breeding coalition develop different but complimenting beak types and the repertoire singing of the younger male magpie that surrogates is the behavioural manifestation of this deep mechanism. This repertoire singing of the surrogating male robin seems to be typical of intergrading members or species that occupy overlapping territories. A similar phenomenon (of beak ssize and song type) is observed among Indian tree pie found in the region (discussed later).

The case of the magpie robins clearly shows that every member in a flock has the capacity to produce song. But why is it that not all the members actualize this capacity and why should only one member overtake the others in the junta? Why is there no smooth intergenerational continuity in song production? The seasonal variation in the song capacity of the surrogating magpie provides some insights into the nature of the social tutor for song and the mode of transmission of intergenational continuity of song. It suggests that the social tutor for song does not actually "teach" a new member the species song (routines), but rather functions symbolically or as a signal. This signal is always negative in modality and produces song suppression. In his first season he does not produce any song post hatching and this coincides with the birth of the surrogating and his female twin. The extensive repertoire singing in the second season coincides with fledging of the breeding male leading to a rejection of the repertoire or song suppression. It is also possible that the reproductive inviability that accompanies this varying song capacity reinforces the song as negative signal. Not only is the surrogating male unable to produce any viable eggs in either season. It appears that the he contributes more to fledging the surrogating member when he produces less song. In the second season he displays inadequacy in his surrogating duties (perhaps causes loss of an egg as the breeding male does in the first season) but accompanied by extensive song. It is also likely that the distance from the nest at which the songs are delivered in each season also has an impact on the song capacity of the new members.

Next from this phenomenal yet plain looking performer we shall turn to another maestro the Tickell's blue flycatcher. Both belong to the same family Muscicapidea.

Tickell's Blue Flycatcher

The Tickell's blue flycatcher (*Muscicapa tickellea*) is an iconic singer that seems to fit the classical mold of song bird. He has steady mastery over song all his lifetime even as the breeding bird. He is not like the iora who is never a tranquil singer unpredictable with his molting problem or the magpie robin who is a mercurial performer and both of which have relinquished their reproductive role for the sake of song. To top it all the Tickell's flycatcher male organizes his tribe in a clan with all the subsequent generations of birds sharing the same territory and expanding it for the purpose.

He is not much of a song learner even though he may have more than one social tutor. He learns his song early as a fledgling and it is a combination of a melodious species typical notes transmitted across generations and a spitting chit chit sounds of non singers in his territory such as jungle prinia and warblers.

To beat it all he has a fine sense of performance and he delivers his song in a rather formal manner, solo of course, perched on a low branch addressing the several directions as he proceeds to sing, twitching his tail up as he spits out the chit sound at the end imitating the little unaccomplished singers in his environment. He starts as early as four in the morning and delivers a concert after ablutions at a leaking tap or a dripping pipeline. Then he resumes again at intervals until nine. Then he may appear again at the same spot at eleven and so on. Thus throughout the day one can hear him sing at last turning in at dusk after another long concert. He

Picture 8: Tickell's Blue Flycatcher (male)

also moves in a circular motion around his territory singing at different locations on his way. This circular movement becomes wider and wider as he completes his season. By a new season's beginning he has fashioned himself a far wider stage area to perform than he started out. By now he also has a couple of offspring fledglings to share his former territory, apprentices yet but who already know their tunes.

In a season it is possible to have three or more males doing their regular rounds of exactly identical song in the territory. In some seasons there can also be a young female not yet entered breeding. The Tickell's flycatcher lead singer then fashions himself in the image of a benign patriarchalist and keeps up steady dominance even over his fledging males. It is not as if he does not allow his fledging apprentices to sing. In fact by the by they learn his song and imitate him almost to the last detail. But the patriarch has an edge he starts his rounds first each morning and takes his turns with his young following him.

The song capacity of females in the flycatcher clan of a territory is of remarkable and shows that it is not true that singing, especially territorial singing,is the prerogative of the male bird only. A female in her first season before laying her eggs can produce phenomenal song like any of the males in her clan. She also becomes part of the clan's song routines and sings the same song as the males suiting them to the occasion and time of day. Post mating the breeding female switches her song routines to night times (spectrograms are given in Chapter One). She also gives up the species typical song and produces songs of other birds that share the territory such as leaf bird and flower pecker. These imitative songs are of single or two notes and are produced in a breathy manner imitative of insects in nighttime. These nighttime songs seem to be directed at monitoring the territory around the nest with the incubating eggs. The loss of typical song and adoption of nighttime territorial mimicry is also accompanied by loss of original plumage colouration from deep blue to a lighter shade. Breeding season falls roughly October to March and the eggs take more than a fortnight to hatch. In her second season she does not produce any song.

Considering the capacity of the female to produce species typical song and then switch over to song learning and sophisticated mimicry it seems appropriate to assign her equal if not more role as singer among her clan. But unfortunately she seems not to qualify for the title of lead singer because she fails to sustain her singing in her second season while her males are able to keep up their species typical sustained singing all through their lives. Still we cannot view this loss of the female's song as a contest rather it seems to be related to her extended reproductive participation.

There are seven other flycatchers sharing the same territory, veriditer (*Muscicapa thalissina*), Asian brown (*Muscicapa duarica*), Asian paradise (*Terpsiphone paradise*), bar winged flycatcher shrike (*Hemipus picatus*), Nilgiri (*Muscicapa albicaudata*), and rusty tailed (*Muscicapa ruficauda*) (canary flycatchers have become extinct or moved out of the territory perhaps owing to the Tickell's presence), but none of them produce any noteworthy song. The Tickell's achieves his melodious song long before all his cousins. The Asian brown manages to overcome the silencing but he achieves no more than a feeble jungle prinia-like "twiii". The veriditer song "twiit twiit" also is no more accomplished than this. The Asian paradise seems to make do with his flamboyant streamer instead and can produce his coucal-like "krrraah" only secretly behind damp rotting trees.

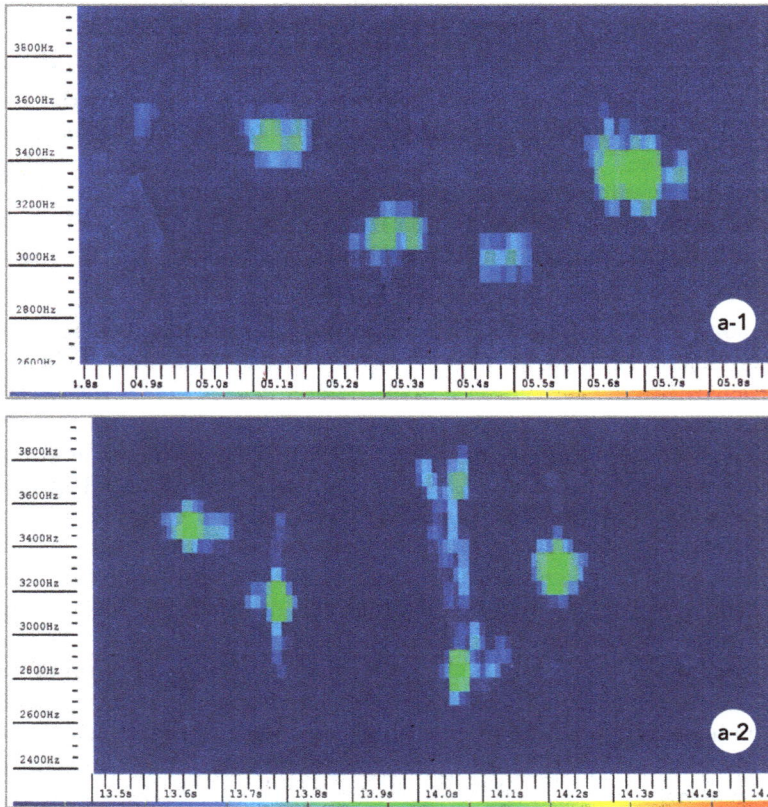

Figures 4a-1 & a-2: Spectrograms of two consecutive sequences of song notes (or two consecutive lines) from a Ticklell's blue flycatcher's morning verse song

What could be the secret of the unvarying song capacity among the members of the Tickell's clan, and the marginal song among the other flycatchers? Now it may be noted that the Tickell's has an unusual reproductive arrangement which is clearly another variation on the breeding coalition among birds I have been observing throughout. An unpaired young male in his first or second season is found to adopt a breeding female (her breeding male may be his twin born out of the same clutch or from a previous season; also see Tickell's family tree in chapter 4). He undertakes reproductive feeding of the breeding female and when the eggs are produced he takes on an active role of surrogate parent uptill hatching. As reward for his contribution he is allowed to pick his female out of the clutch. The role of the actual breeding twin is restricted to overseeing his sibling and female. Even the duty of monitoring the territory is partly taken over by his female.

The Tickell's male always chooses a mate from the subsequent season's, never a fellow member from his own clutch like do birds such as iora, leafbird, bulbuls (passerines) and barbets (non passerines) so on. So the female is always younger than he. Males are always born in pairs as twins and have to bide their turn for a mate since only one female is born each season. A breeding pair can produce either a clutch of three eggs (but eggs are not all laid at once) out of which two are males and one a female or only one egg which invariably turns out to be a female. This difference in breeding capacity I conclude depends on if the female came from a single egg clutch or three-egg clutch. Song capacity also differs similarly. So that every alternate season we can expect a twin male plus a female and in the next only one egg is produced but not necessarily by the same breeding pair but by the second male in the twin.

It is also possible the younger unpaired male also contributes to the nest building. The Tickell's have a finely woven goblet shaped nest with a wide mouth. These nests are attached to the loose ends of young branches at the top rungs of the tree and the privilege of the fledging birds only. Older birds like to reside in tree holes. Since incubation takes more than a fortnight nests may be transferred from place to place overnight for safety. As I recounted on the previous page the breeding female produces suitable nighttime songs which serve to monitor the territory in the dark. The males are quiet during night and sing only in the day.

We find that among the Tickell's unpaired younger males take on the reproductive role usually assigned to females while the breeding female adopts the male role of safeguarding the territory. But the male's participation in the actual reproductive processes is limited to the period of mate selection only. Once he is paired off he will have nothing to do with the raising of the chicks. But the female continues to contribute rest of her lifetime and stays aloof from song after her first season. In all probability the leading male singer in the clan forgoes all involvement in any kind of reproductive activity after his first season and devotes his entire time to song. After all there are always younger males in his ever increasing clan. Our patriarch seems to choose a role larger than mere spouse and exists as the head of his clan.

Pictures 9a, b & c: Tickell's female, fledgling & nest

It must be pointed out that this unusual breeding arrangement strongly points to the unlikelihood of the female having an upper hand in mate selection. It seems that among the Tickell's it is the male's prerogative to choose his mate and his surrogate parent role serves to bind his sexual right over his female.This sort of reproductive praxis I suggest cuts off the female from the males of other flycatchers sharing the territory (isolating mechanism). Perhaps her ability to sing the species typical song in her first season and her role as territorial nighttime guardian also strengthens her belonging to the clan; her ability to sing confirms her affiliation.

Since all the flycatchers belong to the same tribe and share the same niche we would expect the same kind of reproductive arrangement among them. But strangely other flycatchers that share his territory do not display this kind of reproductive arrangement. In fact among the Asian brown we find that the female dominates the reproductive processes. This is true of the Asian paradise whose female loses her beautiful tail streamers permanently on undertaking breeding. I presume the same kind of reproductive arrangement exists with the veriditer also. Only the Tickell's has the prerogative. All these flycatcher species are well matched in size, morphology, foraging and nesting habits.

An intriguing fact is that the young Tickell's male indulges in hawking on the ground for grubs when foraging for his unfledged female. This I feel is an important domineering activity. The females glean discreetly and don't hawk. Hawking obviously requires mastery something like how a raptor catches preys by swooping from above to the prey on the ground. The young Tickell's bird skids on his legs on the ground when he sees a grub and catches it in mid air as it flies off then returns back to his perch on a low branch. He also produces a whistling vocalization when he does this. Now among the other flycatchers in the territory this type of foraging is done by the breeding female. The males forage on tree top and never come to the ground.

Pictures 10a & b: Asian Brown Flycatcher displays sexual dimorphism

The flycatchers in the region are one of the largest and most diverse tribe of birds of their family that includes also chats and thrushes. In the habitat of the Tickell's blue flycatcher we find the magpie robin who is a member of the chat family and a removed cousin of the flycatchers.Other members of the family such as pied bushchat and Indian robin inhabit a territory just outside of that of the Tickell's. Pied bush chat does not produce any elaborate song but have highly expressive vocalizations that are highly communicative. Among them the chicks are raised according to their sex by the male or female respectively.

What could be the underlying socio-biological rationale that explains the complete suppression of song of the other flycatchers and the consequent structures of dominance? How did the Tickell's come to establish his prerogative over his cousins that share the same niche as he? Why does he not resolve it in the same way as the leaf birds and orioles-tree pie do (see sections below)? Literature on Asian Brown is very uncertain about their status and claims that their breeding range may be limited to the north eastern Himalayas (vol 7 Ali & Ripley, 2001) and they only winter in west coast Southern India. The veriditer is also given the same range as Asian Brown and are found to breed in the Himalayas but winter in Southern India. The Tickell's is thought to breed in whole of southern India and may be found up to the slopes of the Himalayas. I suggest these accounts are incorrect; all the three flycatchers have the same range the ambiguity in field observation is due to the dominance of the Tickell's over the other flycatcher species.

The reproductive arrangements among the various flycatchers and other behavioural adaptations strongly elicit the hypothesis that speciation events and perpetuation of each species cannot be understood purely at the level of individual species but should be viewed as a cluster event. Species within a cluster must be involved in a symbiotic society in which there is a super cooperation between the members on every count including intergenerational continuity. It is my hunch that the Tickell's plays the role of chieftain, not merely head of his own clan. A question will arise of course why should the Tickell's have the prerogative? Why should he dominate? Perhaps the answer lies in the genetic makeup and evolution of the Tickell's and other flycatchers. It is possible that he is a true scion of the old world flycatchers while others evolved later. The dominance of the Tickell's is genetically sanctioned and he is in the position of authority to safeguard the flycatcher diversity.

In this regard, in addition to clustered speciation, it seems the type of reproductive arrangement is crucial to speciation. Our Tickell's therefore enforces that his cousins do not share his reproductive arrangement and have a different sex relations, which ensures their species distinction.By reproductive arrangement I do not mean ban against interbreeding but something beyond that. The idea of reproductive arrangement or what I have called praxis has been central to this book and my field observation in the Bhadra region has thrown up several new heretofore unrecorded dimensions of this praxis among birds.Thus among the Tickell's mates are chosen cross generationally, never from the same clutch; among Asian brown mates belong to the same clutch; among the magpie robins the mate belong to different interbreeding populations, so on.

Leaf birds

The exotic leaf bird (Chloropsis) is one of the oriental birds much sought after as a pet because of their capacity for vocalizations. The market for them is so brisk that bird lovers fear that they are nearing extinction since they are endemic to this region and do not breed outside their indo Malayan range. Thus this colourful expressive species seem to perpetuate the image of Asia as the exotic carefree orient. These birds like to nest in flowering trees because this means they have ready food, insect and nectar, and later fruits resource at hand. This factor and the involvement of both male and female in fledging make them vulnerable and easy to catch. Furthermore they are gregarious, intelligent and highly responsive even to humans. Forest edges with flowering trees such as tree jasmine and flame of the forest are enlivened by the variety of expressive sounds they produce.

Picture 11: Golden Fronted Leaf Bird

It appears that these birds have developed several mechanisms to combat predators, poachers and possibly rivals. One such device is the keeping of mock nesting routine. As a rule most birds have several nests in different places at the same time this helps to ward off predators and protect from destitution. Having several nests at the same time and following a mock nesting routine as a decoy enables the leaf birds to ward off predators.The mock nest is built on tall tree tops on weak twigs like the real nests. The second mechanism is flocking of these birds in mixed sub species all-male or sex segregated parties in their territory. The males of golden fronted and blue winged leaf birds are found to flock and forage together on the same tree, male member of one species feigning as female of the other species. A complementary mechanism is the displays of the non breeding female. During off season she exhibit individual displays like rolling and hanging from tree. Only a close observer who knows the songs of these birds will recognize them as different species. A further obstacle to recognition is the reverse song type matching that is practiced between them. By reverse song type matching I mean a situation where one species produces the song of the other in the reverse order of notes. Camouflaging seems to be the most important aspect of these mechanisms. But certain ambiguity exists in this assignation because in Ornithological literature such mechanisms as sex segregated flocking have been viewed as species isolating mechanism. Perhaps it is important to note here that it is possible that they primarily evolved as devices against interbreeding which have become suitably extended to camouflaging in response to human threat.

In the region there are four species of leaf birds: The subspecies taxanomy that I have used in the following is derived from Ali's inventory (vol 6, Ali & Ripley, 2001). But he lists only three subspecies for the entire South Indian region and only two for the Bhadra. It is possible that a new taxonomy is required for any accurate reference. I make this caveat because the morphology of particularly one species does not match any of Ali's description and I am forced to refer to this particular species as a new southern race (NSR). There is no mistaking the presence of the four distinctive species because each produces distinctive song in addition to distinctive morphology and do not interbreed (breeding seasons are different).

Two of the species the blue winged (*Chloropsis cochinchinensis*), and golden fronted frontalis (*C. aurifrons frontalis*) occupy the hilly deeply forested epicenter area, with tall perennial trees. They share the same territory; only the gradient of their range being in opposing directions: south west to north east and north east to south west, respectively. Between them they engage in reverse song type matching. Peak breeding season is April to August and March to July, respectively. The same territory is shared for foraging as well as nesting. There may be turn taking in foraging. But nesting is undertaken on different varieties of trees well separated even if in the same territory. They are visible during the breeding season and lie low once the fledglings are out of the nest and ready to take wings. This is also true for the vocalizations they produce which are maximal in season only.

Pictures 12a & b: Blue Winged Leaf Bird (male) & nest

The two other leaf birds insularis and NSR on the other hand occupy different ground types that circumference the epicenter. The NSR occupies open grassy fields (scrubland) bordered sparsely with flowering perennial shrubs and trees. The insularis occupies man made forests on the plains that border the scrublands. These two are not found together, *i.e.,* they do not intergrade with the other two or among themselves, even though have comparable vocalizations between them. Peak breeding seasons are June–October and December to March (insularis). NSR builds open triangular cup shaped nest using very fine grasses in the middle rung forks of trees, which is unusual for leafbirds. The others build crude closed egg shaped nests with plant stems and twigs, generally placed on overhanging canopy of tall flowering trees. The cochinchinensis nest can be found hidden in the middle rung forks like the NSR. But they have the same closed egg shape as of the frontalis nests, but are made with slender plant stems finely woven.

Pictures 13a & b: Frontalis fledgling (female) & nest

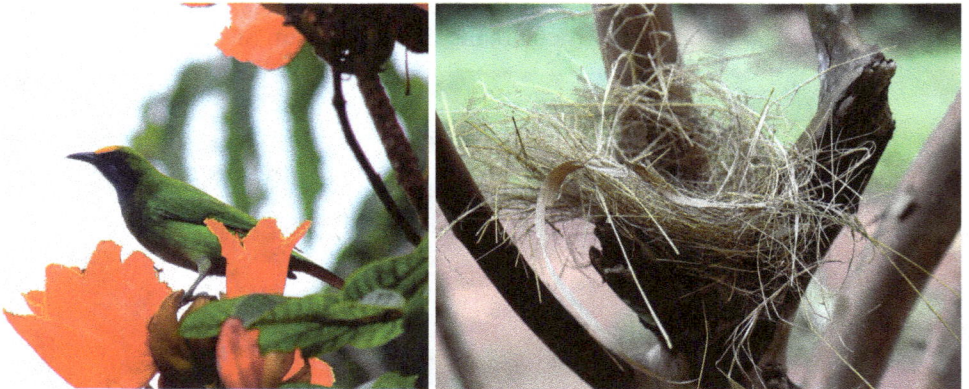

Pictures 14 &15: NSR and nest

Between the blue winged and golden fronted frontalis leaf birds there is a clever reverse song type matching occurrence. Both of them are found together especially in the months of November to May in small sex segregated parties. It is the non breeding male birds that indulge in the song matching. The typical song of the blue winged leaf bird is "wee" repeated at intervals. But the singing bird very seldom produces its own song purely. Sound imitative of other creatures in the region such as squirrels, chup chup or the drongo's pip pip are tagged on to the typical song to produce a longer variation such as wee wee chip chip or wee wee chup chup. The frontalis sharing the same territory produces a reverse pattern chip pip wee or pip chup wee. These reverse variants also can have other sounds strung along such as that of myna, parrots, robins and bulbuls which mark out the identity of the bird as originating from a different territory.The reverse song matching can be accompanied by ventroquilism wherein a frontalis male hustles the male of his kin cochinchinensis using him as the ventroquilist dummy and as cover. Such song throwing seems to be used for scenario testing and can have sounds picked up from the surrounding rather than the species vocalization. In such a pairing the cochinchinensis male on the other hand feigns to be the shy female. But when accompanied by his female he will pretend to be a fledgling and boldly call out.

Table 2: Two different types of reverse song type matching recorded in the Bhadra area

Reverse song type matching -1 Chlorospsis		Reverse song type matching -2	
Cochinchinensis	**Frontalis**	**Purple rumped sunbird**	**Purple sunbird**
Wee ´wee´ weeweechipchip weeweechupchup chipchipchipchip pip	chip pip wee? pip chup wee? Pipchuppipchuppip crack crarrara wee? wee?	i) cheleep cheelep ii) che che ple che ple pe pe pe che che ple che ple...	i) cheeple, cheeple ii) ple ple che ple che... iii) cheep
Insularis: wee wee wee wee twwieue druwitu druwitu wee wee wee wee pip pip pippip pippippip	**NSR leafbird** twee twee twee twee wieue twituwi twitwitu kii kiihm kii kiki kiikm kii		

Note: Among leaf birds there is always only one singing bird in the flock. Among sunbirds only males display reverse song type matching.

The third and fourth species insularis and NSR leaf birds do not intergrade among themselves or with the other two and choose to nest in the adjoining territories that circumference the epicenter in the area under investigation in this book. Vocalizations of the NSR and insualris species can be heard maximum in August- September and December-March respectively. It is possible that analysis of the NSR (and insularis) songs will reveal them to be the most elaborate among them. A whistling longer variation"twee wee wieue" is produced by the NSR interspersed with imitations of young shikra calls "kii kiihm kii kiki kiikm kii" or that of minivet "twituwitwitwitu" both birds that nest in its territory. The insularis produced a song "wee twieue" accompanied with "pip pip" of the drongos that was recognizably in the reverse order of note of the NSR. It is likely the gradient of expansion of the insularis is opposite of NSR, south west to north east and north east to south west, respectively.

Before we examine the song matching phenomenon more closely it is important to note that among the leaf birds even though they are found in flocks only one bird produces the song. The bird that sings is the non breeding male. The breeding pairs among all the subspecies seem to have equal participation in the reproduction and are not visible during the season. Song is undertaken by a non breeding male during this time. Non breeding females on the other hand are found to produce identical vocalizations in the off season. But they are more likely to exhibit displays like rolling and hanging from tree rather than song during off season. On such occasions the female is found singly and more over the birds indulge in sexual segregated grouping.

How do we understand the differences in the songs and song matching among the species of leaf birds? Why should there be reverse song type matching between only the blue winged and frontalis and insualris and NSR? Why is there no evidence of song matching between frontalis and NSR? Why is the song of the NSR (and insualris) even though similar but more elaborate than the blue winged and frontalis? Why are the breeding seasons of the blue winged and frontalis overlapping but that of the NSR (and insularis) falls in a totally different time of the year? Why do the NSR and insularis inhabit the margins of the territory and different ground types? Why do the leaf birds not engage in any song matching or contest with other bird species in the territory, such as Tickell's flycatcher, mynas, parrots and hornbills, so on?

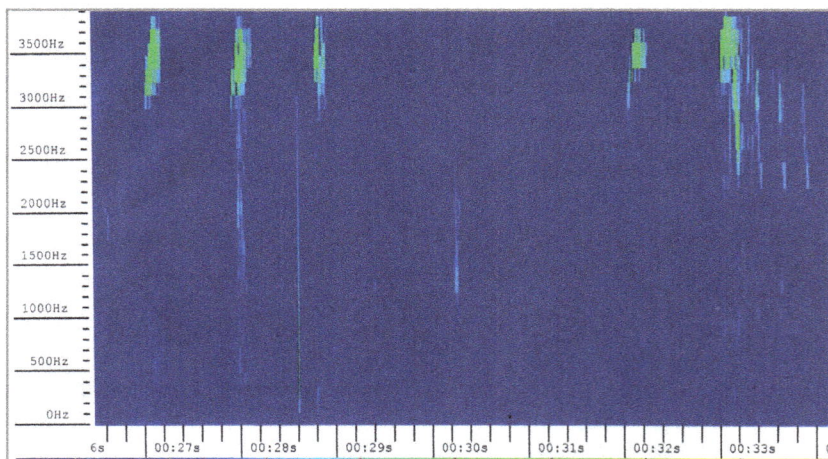

Figure 5: Showing a sequence of notes ("wee, wee, wee, wee, weechippippip") from the leaf bird song

In chapter two I have proposed that songs of birds must be mechanism of species cohesion and index to the genetic relations in a bird population. The similarities and differences in the song patterns of the four species of leaf bird strongly elicit this conclusion. The song distribution among them and the consequent reverse song matching indicate that the cochinchinensis and frontalis share the same relation between them as do the insularis and the NSR. The more elaborate song of the NSR and insularis than the other two also must be a key to the kinship between them.Field literature documents that leaf bird stock migrated from north eastern India to the Indian peninsula and underwent subspecies adaptations. But a close analysis of my field observation has raised some challenges about speciation and the genetic relation among these birds. It highlights the problem of understanding the ecological speciation terms like allopatry and sympatry.

There is no doubt about the existence of the four species because each is distinguished by its own different song, type of nest and breeding season. They also show morphological differences such as size of body, shape of the beak and claws and in colouration. For instance the NSR has a lighter blue chin patch without the black border. Its beak is finer and claws longer I suggest with which to pull out the clumps of grass that grow hardy in the scrub land plains for nest building. The insularis is the smallest and darkest green, morphologically similar to frontalis but far smaller in size. It lives in montane forest of the plains that border the scrubland. These montane forests have thick forests of tall lush evergreen trees like honge. The cochinchinensis does not have the orange crown but only a deep yellow. It is of same body size as the frontalis but with finer beak. It is possible it lost the species typical orange front in camouflage response to colours that predominate in its habitat.

The diversity of leaf bird species in the Bhadra region seems to foreground some very interesting clues to species existence and subspeciation. David Lack in his book Ecological Isolation in Birds (Lack, 1971) tried to solve the puzzle of Darwin's finches. He came up with the idea of competitive selection as the mechanism for subspeciation in island biogeography. According to him a later generation of migrating bird can replace a previously migrated stock and develop better adapted body types suitable to the niche available. Underlying his hypothesis is the inavialabity of diversity of habitat in island ecology. The diversity of leaf birds in this region present to us the reverse scenario where a diversity of habitat is available and is made use of by the same species by diversivfying in adaptations in subsequent generations. Competitive selection in this case has not replaced an older generation with new but diverted the second generation to share micro habitat differences across their range.

Possible routes of migration that lead to ecological divergences
among a same stock/generation of leaf birds

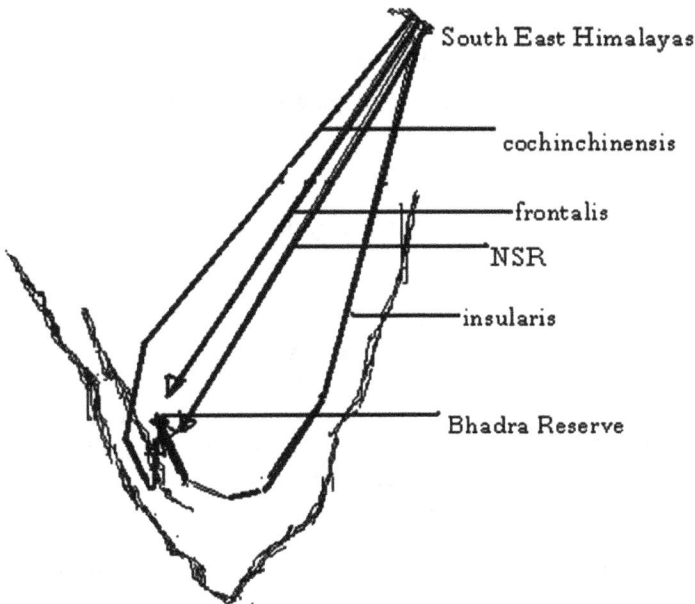

Figure One: A map of the possible routes of migration that caused ecological divergences among a same stock/generation of leaf birds

The cochinchinensis and frontalis have a range along the western ghat mountains expanding in mutually opposing directions, The insularis are inland birds expanding inwards, perhaps across the Deccan Plateau. The NSR have a range in the foot hill plains of the Ghats expanding south west and then perhaps turning south east across the range that links the eastern with the wester ghats. It appears that the intergrading species cochinchinensis and frontalis must have speciated in response to similar kind of habitat (hilly evergreen forest) even though in different regions of their range and the Bhadra region must be their secondary contact zone (therefore are geneticaly closest). Like wise the NSR and the insularis (plains, montane). But it is possible that the frontalis and the NSR (cochinchinensis and insularis) speciated competitively in response to microhabitat differences with in a same region but migrating in successive generations (possibly to be viewed as parapatry but turns out to be only in relationship perhaps owing to speciation in successive migrations) but which resulted in non intergrading speciation, so that genetically they are furthest between them (divergent adaptation). No song matching is found between them. But one of these pairs has developed a more elaborate song (comparable to among the three bulbul species found here). On the other side the close reverse song type matching between the frontalis and cochinchinensis (NSR and insularis) indicates their genetic nearness derived from having originated in similar allopatric conditions but in different regions, but sympatric and intergrading in their secondary contact zone. My suggestion is that these ambiguities indicate that ecological speciation categories are insufficient to understand the genetics of speciation (see Appendix 4). And song matching gives a clue to speciation relation as index to the underlying genetic make up rather than by the mode of speciation.

Racket Tailed & King Drongos

The drongos are a family of passerines commonly sighted all over India. References to them in Indian classical literature suggest that they must have been common even in ancient times. Going by the field literature six tribes of drongos may be listed for the region: they occupy the same territory, appear to share the same niche for nesting and feeding. Also they may not be distinguished by plumage. It is likely this classification needs a revision. Vocalization records in the region confirm only two species king drongo (Dicrurus macrocercus) and racket tailed (Dicrurus paradisius). The existence of a third species is confirmed by its distinctive spangles (Dicrurus hottentottus). It is possible that all other species listed such as ashy and bronzed (D luecophaeus and D.aenues) drongo in field guides are merely either the racket tailed or the spangled drongo in post- breeding plumage. The breeding members (both male and female) lose their rackets (spangles) during incubation or later due to molting and can seem to be different species. Drongos are typically found in mixed flocks, with birds like tree pie, oriole, malakoha scimitar and other babblers, Asian paradise flycatcher. When in such a mixed flock an orchestra or a medley of displays and songs can be had. The drongos are not good mimics and are selective in imitating other species bird songs. The king drongo in the territory produces imitations of the Shikra juvenile member during post breeding time.

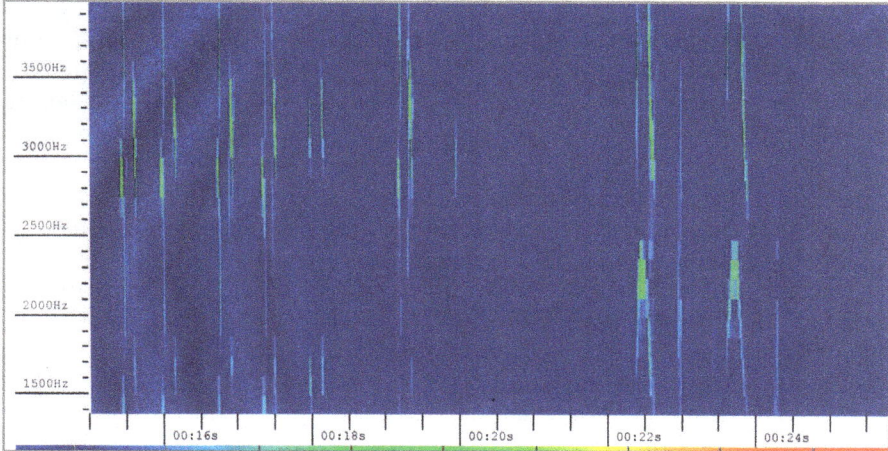

Figure 6: A note sequence ("draan (not shown in figure), kiki, kiia") from the king drongo's post breeding song

Among the drongos the racket tailed is the most impressive not merely for its long twin rackets and imposing crest but for the elaborate song. The peak song can run up to 52 lines made up of 2 to 20 notes delivered with great ceremony and pomp from the top of a wintering tree. It is a local resident bird in the Bhadra region and may be found in sizeable flocks of four to six members throughout the year frolicking among tree canopies and under hanging boughs in light forest thickets. But newly fledged members diversify by migrating to other forests as soon as fledged leaving behind their parents who have lost their rackets during breeding. The rackets enable the younger birds to fly swiftly in the heavily moisture laden air during high winter. The rackets are a pair of tail feathers with elongated shafts. The deeply forked tail of the bird and the rackets seem to have the same function as tail fin in fish and can even break light ice formations in the winter air. They can fly upto 300 km or more. Those that have lost the racket cannot keep pace. The breeding season falls in monsoon starting July to November.

Pictures 16a, b & c: Racket Tailed Drongo & Close up of the rackets and crest

Even though no subspecies are listed for the racket tailed it is my suggestion that they diversify clinally correlating to the variation in nesting altitude among different populations. In the field literature on the racket tailed two species are identified greater and lesser racket tailed by the size of the crest and length of the tail. Ali (Ali & Ripley, 2002) reports the greater racket tailed for this region. My field observation indicates that the length of the racket and crest size vary with the altitude of the nesting sites of the species members. Members that nest on absolutely tall trees in hilly regions have extremely long tail feathers with a deep fork as well as rackets. Those that nest on medium sized trees on hilly slopes have less deep tail forks and shorter rackets. Those of the plains have the shortest tails and rackets. This variation in tail and racket correlates with the ability of the birds to fly in different altitudes in moist wet climate. In addition to this the shape of the head and beak also varies. Those with the longest rackets and deepest forks have forehead that appears angular and flattened, while others have more rounded. The beak size and shape seem to correlate with their nesting habit, not unlike the magpie robin beak variation.

The nesting habit of the racket tailed is of great interst. Typically they nest in hollow forks of large sized trees produced by forest fires or in rotting trees some ten feet above the ground. The tree hole has several openings (something like a hollow square) and is cup shaped at the bottom and unlined with any padding. These several openings could be to accommodate the rackets of the fledglings. Such a nest is large enough to hold a good size clutch. Alternatively the bird can improvise a nest by prising open the bark of a tree in a rectangular boat shape and placing it on overhanging tree canopy at the absolute top of trees. Such a nest is used for a smaller sized (perhaps a two- egg) clutch. It is my suggestion that these two different nest types are used for differential allopatric rearing of chicks to become breeding and singing. Those reared in the tree holes molt easily and lose the tail streamers without any harm to the life of the bird, but if raised in the tree top bark nest molting is usally fatal to the bird. It appears

Pictures 17a & b: Racket tailed at the begining of post breeding molt and at the end

that tree hole bred members have different beak shape from tree-top bark nesters. Also perhaps the drongo nesting habit is adaptation to maximize the uses of forest fires that are frequent and seasonal in the natural setting.

The drongos are sexually unmarked it is near impossible to make out the male and female. Among the racket tailed all the birds in a flock have the rackets and the crest season initially, but as the season proceeds some of them lose the tail streamers. It is possible the crests of all members are foreshortened during this time. But the members that incubate the nest eggs lose their rackets completely and seem never to regain them. At the end of the breeding season in November one can see a good numbers of racket tailed drongos without the racket. They tend to flock together and appear as a new species. But there are always members with the racket through out this season. Such members eventually lose their rackets due to aging.

It is possible racket tailed are colony breeders with several breeding members attached to a leading male. A large sized clutch is distributed in several nests with the members playing the breeding role contributing to the fledging. The leading male plays the monitoring role only. He therefore does not lose his rackets and is the singing bird among them.The end of a successful season is heralded rather dramatically by a pair of birds one with the racket and the other without it making an appearance together as part of a song routine. On these appearances the bird

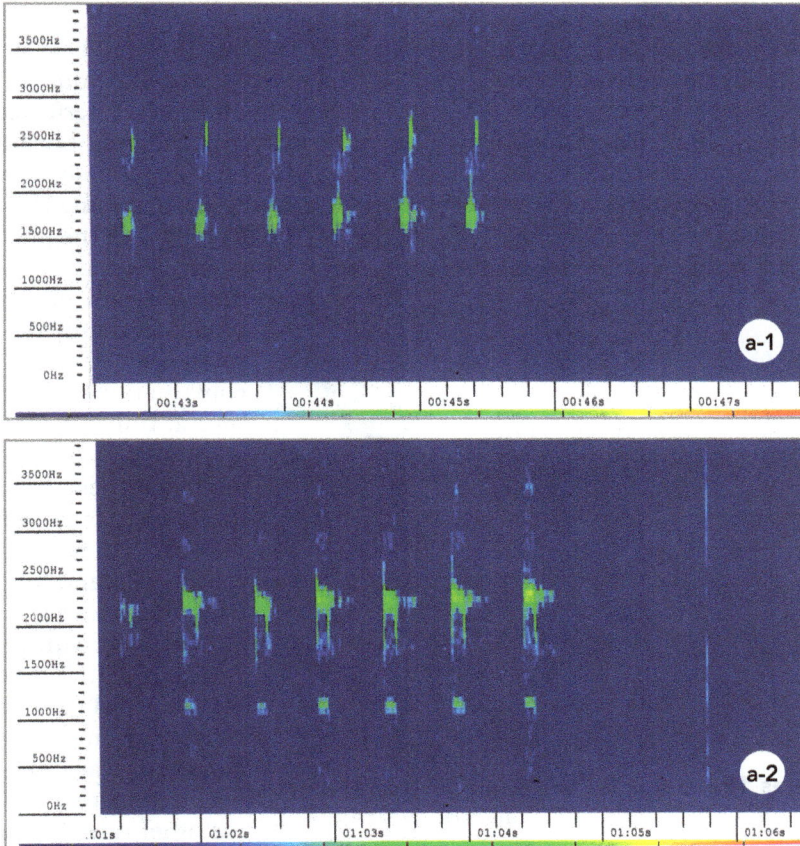

Figures 7a-1 & a-2: Two distinctive & successive lines ("droyn, troyn") from the racket tailed drongo's post breeding song

with the racket delivers his song from the top of wintering bare trees,while the bird without the racket merely sits on the same tree as he at a lower level.

In comparison with other drongos the racket tailed drongo has a more elaborate song as well as song routine. For one they produce distinctive vocalizations varying with the season and status of the bird. In February- March one is bound to hear and see young pairs (very likely the breeding pair) in their first season playing trapeze (like in a circus) on tree tops producing vocalizations similar to the king drongo. It sounds like a whistle followed by the sound "pip, pip." At the start of the breeding season in July to August one is bound to hear another type of song, a kind of pre syllabic velar nasal sound like a brook running or a water musical instrument. Peak singing can be had in November when the chicks are newly fledged. Both of these songs are delivered by the monitoring member. This singing is delivered from tree top with much ceremony and with an elaborate song routine spread across seven continuous days. The syllables and sound notes of this song are a combination of the juvenile vocalizations and the pre syllabic utterances of the season initial. But the versification of these sounds makes it elaborate and complexly structured.

The season initial July-August song of the monitoring bird is sung intermittently throughout the day especially mid mornings from behind foliage of trees while the bird swings from boughs. The singer is alone while performing; he is shy of being sighted and shuns visibility. The peak singing in November is carried on early in the mornings after sunrise and repeated over seven continuous days and can run up to 15 minutes each time. On some of the days it is accompanied by a mate. Spreading across seven days, this song begins on the first day with the presyllabic velar nasal sound like the initial notes of a bugle or conch. Then as days pass it develops into fully formed syllables "droyn, troyn" produced in various combinations and repeated several times ending in an additional sound note "ta" and "ka."These verse lines are followed by lines entirely formed of the sound "pip" repeated several times with slight variations in scale interspersed with the "ka" sound. Imitative references (mimicry of other bird songs but in one's own dialect) to other birds in the nesting territory makes up the other lines in the song.

Pictures 18a, b & c: Racket tailed clinal diversity based on racket length, tail & body size

This verse song runs up to 52 distinctive lines delivered for 15 minutes continuously. The elaborate song of the racket tailed has a ritualistic quality and possibly is a mode of inducting the newly born chicks as stake holders in the territory's community.

In Table Three I have displayed the song data for two drongo species. The individual sound notes and the vocalization patterns of the racket tailed is so elaborate that I am unable to present all of the transcribed data here. Since the peak song runs up to 52 lines each made up of 2 to 20 notes. The full song is delivered for up to seven continuous days. Here I have displayed representative lines only.It must be noted that in all the accounts of song I give in this book I have distinguished between what may be called as a complete song and repetition of the song. Birds deliver the complete song repeatedly in most instances. So the tables will not indicate the total amount of vocalizations of the birds but only the minimal units.

Table 3: A Comparison of Vocalizations of Racket tailed and King Drongo

Status of bird	Racket tailed	King drongo
Immature adult	Whistle + pip pip	pip pip
In Breeding	pre syllabic version of "droyn"	draan draan pip pip
Post fledging	droyn troyn droyn troyn droyn	draan
(first five lines only)	troyn droyn troyn	kiki kiki kiki kiki
	droyn troyn droyn troyn droyn	kiki kiki kiki kiki
	troyn droyn troyn droyn troyn	kiki kiki kiki kiki
	droyn troyn droyn troyn troyn	draan
	droyntroyn droyn troyn droyn ta	kiki kiki kiki kikki kiki
	droyn troyn droyn troyn droyn	kiki kik kiki
	troyn	kia kia
	ka ha ip pip ip nam na	kia kia
		draan
		kiki kia

Like most passerines (iora, babbler and purple rumped sunbird are the exception) for whom data have been presented in this book the vocalizations of both the racket tailed and king drongo has a component which is species typical and a component that is made up of imitation of other bird songs. A good comparison may be found in the species typical song note even though it appears to be produced in two different dialects. In the racket tailed song there is a note (troyn) which is not present in the king drongo but is not mimicry of other birds (ka, kiki, kiow) in its habitat. In addition the imitation sounds in the racket tailed are hybridized to a greater extent, in the king drongo song they still retain their original identity. But song pattern of the racket tailed is entirely different and far more elaborate both in its species typical component and imitative component.

Perhaps it would not be wrong to draw a generalization and draw a similarity with the song patterns between two species of leaf birds Cochinchinensis and insularis/NSR and frontalis. Another comparison may be found between three species of bulbuls. In all these examples we find lateral member species of a family clade adopting a more elaborate variation of the species specific song by generating new notes, even while retaining the other sound notes. Perhaps this can lead us to a generalization that such song elaboration reflects cladogenesis rather than ecological speciation. Below I discuss the song of the hooded oriole and find that they produce a comparable song to that of tree pies their cousins but in a different dialect. This case study also confirms the above generalization. The three species are grouped under one polytypic family corvidea.

Hooded Oriole and IndianTree Pie

The orioles are a family of birds grouped under corvidea of which the rufous tree pie also is a member. The literature lists some thirty species of orioles; only three of them were sighted in the Bhadra area; Black hooded (*Oriolus xanthornus*), black naped (*Oriolus chinesis*) and Eurasian golden oriole (*Oriolus oriolus*). It is likely that all of them breed in the region, and have comparable breeding seasons falling between October-November and March-April even though the literature lists them as merely winter visitors. Large sized sex segregated flocks of the latter two species (six or more) become visible in April. It is possible that this confusion about their resident status is because they do not nest in the same location in all the seasons and migrate locally to adjacent groves. After May all the species lie low (are not visible) probably giving a chance for the fledglings to become mature adults. It is unlikely they migrate a great distance in the interim since the chicks are just out and still have to find their bearings. The sex segregation among orioles is similar to that found among the Indian cuckoos. Among the Eurasian golden oriole species the older females appear to dominate over the younger in such segregated groups. The black hooded (Oriolus xanthornus) remain segregated most of the year, the opposite sexes pairing off only at the beginning of the breeding season December to May. In the following I give data for the black hooded oriole.

Pictures 19 & 20: Hooded Oriole (male& female)

The male black hooded oriole has a most remarkable song that does not reach its full tonal variations until the last week of April-May even then he sings it briefly spanning about five-six days. He begins his song on the first day with a crow-like "Krau"starting early morning and intermittently during the day. Then it expands to a low vibrating (like that of an insect) singing "krayaru" at dawn. By late afternoon this has developed into a full song "krakra k kra k kru" (a hollow U followed after a glottal K). This song is subsequently sung with a variation "kra$_{ss}$ukra$_{ss}$ukukrakukru" for another day in the afternoon. On the fifth day at last the final perfect and longest version is produced "krau krau kruwu kruwu ukru ukru" from early dawn to sun up. Each of these days his developing song is rendered from under canopy of different tree top foliage in the territory. In the days that followed the bird failed to sustain his singing and reverted to the earlier day versions of song after which he could not be sighted any more. In all this the female produced only a one-note vocalization a nasel crow-like "kraak"that never developed to a song like the male's. Her one note vocalizations were accompanied by bodily displays of rolling and somersaulting, typical poses of passerine birds in rut. But this was only season initially in November-December and season end in May. Between these months she is totally invisible. She can be sighted again from May but she does not produce any song. Instead she displays cuckoo female like behavior, such as swinging on branches and feeding on berries. She also goes into a molt and becomes a brighter yellow. The molt may be caused by the feeding on berries to enable them to produce saliva with which to strengthen the nest.

The hooded orioles do not sit on the egg to incubate it but use the heat of the sun. Cradle shaped nests are slung hammock like in the open foliages of tall trees for the purpose. The nests are built out of either discarded or stolen nests of warblers and flycatchers (paradise flycatcher's nests are a favourite) since the orioles seem not to have the beak structure to weave these finely woven nests. The finely woven nests of smaller sized birds are used as starting material and merely buttressed to the twigs of the open branches. The male hooded oriole has a finer beak than his female and consequently takes over once the eggs are hatched. It is his role to feed and fledge them. The method used for hatching enables the eggs to hatch in merely three days time in my observation by which time the female starts to go in to post breeding molt.

The timings of the songs of the sexes and their visibility-invisibility in the territory seem to be of significance: I suggest it coincides with their respective involvement in the reproductive process. The female starts early and moves out of the picture early. Her one note vocalizations inaugurate the season (Nov-Dec) but she never produces full song. The male starts late (April-May) and makes his stage appearance only after the eggs are laid. He in fact does not reach his full song capacity until then. He sings briefly his peak song and then becomes invisible. Again this seems to suggest the impossibility of song capacity playing a role in mate selection. The hooded oriole song demands for an alternative explanation for bird song just as the Tickell's did in the previous section.

In the following I also present data on Indian or rufous tree pies (Dendrocitta vagabunda) that are also resident breeders in the same territory as the hooded orioles and belong to the same family corvidea. The song routines are completely different in terms of sex distribution, quantity and duration, even though the structure of the notes is comparable. There is no occurrence of confrontational song matching between them, meaning there is no evidence of interaction between them. The tree pies have a song for each season, a pre breeding, in breeding and post breeding. A young treepie (D V pallida) whose sexuality is not yet confirmed sounds a polyphonous "kukra kma" (for D V parvula "kukra ka?") This becomes sex differentiated with the male and female tree pies developing their own sexed versions of the song each; in breeding male kukra kukru kaakaa or female kukra kukru maaku. Post breeding the song is sung in a call and answer duet mode between the sexes - call Kra$_s$gudu$_s$kra$_s$gudu$_s$ answer krang.

Picture 21: A pair of Indian Tree Pie post breeding

Adult tree pies are sexually indistinguishable but for their song differences, whereas the hooded orioles are totally sex differentiated in song and morphology. Tree pie breeding season spans from April to November, and the songs are spread across these months. So it appears that tree pies are more heard than the hooded oriole that produces his song only for not more than a week. The breeding black hooded cannot or will not sustain his singing and is a rare bird compared to his cousin tree pie. It is possible that this difference in song distribution is due to different type of reproductive arrangement adopted by the two species. Among the hooded oriole the breeding male contributes a great deal to the fledging and cannot sustain his song. Among the Indian tree pie (D.v. pallida) my observations indicate that fledging of the chicks is in the care of an older pair and the younger breeding pair does not contribute further than the initial stage.

In ornithological literature on the Indian tree pies they are listed as a family with as many as five geographical varieties. Salim Ali in his Birds of India and Pakistan allots the five subspecies to five different geographical areas covering up till the foot of the Himalayas. Two species are assigned to the Western Ghats, pallida (north) and parvula (south). In his account of the distribution of the pallida says "Intergrades into bristoli in the north and west (Sind and Baluchistan), into vagabunda on the east [Andhra], and into parvula in the south [Kerala]. The populations are contiguous and their differences entirely clinal, therefore no more than an approximation of their range is possible or warranted (Ali & Ripley, 2001, vol 3, p. 217)." Ali developed the subspecies classification on average mean body type and length of the tail. The tree pies plumage morphology is invariant among all the so assigned sub species.

Based on his subspecies distribution mapping it is tempting to attribute the two varieties of tree pies I sighted in the Bhadra region as *Dendrocitta vagabunda* pallida and D V. parvula. The two varieties are distinguished by different territories that overlap in the area under observation. That is the parvula occupies a territory that encircles that of the pallida. The same are used for breeding. The two display subtle morphological differences, the parvula breeding male being visibly larger in size, darker tawny, with a sootier hood, beak is larger *etc*. They also display remarkable differences in song and breeding seasons; that of the pallida falls from April to July and that of the parvula from August to December. But it is my suggestion that one (parvula) of these is a derivative subspecies of the other (pallida) and that the two are partially interbreeding; that is they interbreed intermittently. Therefore Ali's classification is not strictly corroborated. There are other evidences for this discrepancy. One, in this territory it is the parvula that intergrades into the pallida territory in every alternative breeding season even forcing the other to find another nearby territory for their own use in that season. In Ali's account it is the pallida that is intergrading. Two, in his description the parvula is the smallest while the bird I

Picture 22: Indian Tree Pie

have assigned this name is bigger than pallida. Thirdly Ali's account of parvula's breeding season does not match with mine August to December.

It appears that these two populations are not merely subspecies but together make up a breeding system very likely to maintain species cohesion. They are intermittently interbreeding and cyclically replace each other in the two habitats and in that season the one coming to occupy the hill slope elects a breeding member (either a male or a female) from the other population that gives up the hill slope electing to nest on the plains (this can be visualized as a pull and push mechanism). So that cyclically those on the hills slope come to occupy the plains and vice versa. By this two distinctive populations are safeguarded through out and morphological differences in body sizes including beak sizes *etc.* is evident between them, with the members on the hill slope being smaller in optimum size than those on the plains. It appears that breeding continuously on the hill slopes leads to incremental decrease in beak and body size, breeding continuously on the plains leads to the opposite morphology both of which are inviable after a few generations. It perhaps also causes the loss of species identity. Therefore the tree pies maintain two populations and interbreed intermittently. Like the magpie robins and sunbirds (see below) the tree pies' unique breeding system points to the plasticity of genetics of species and raises interesting questions about what is meant by adaptation. It also throws light on subspecies existence.

An intriguing aspect of this mating system is the distinctive vocalizing behaviours between the two populations. Distinctive songs are produced only intermitently when the event of interbreeding takes place. In the other seasons the two populations have the same song (perhaps there may be a gradient of change in quality of song notes but only in small measures). These two songs vary greatly in their types, the simpler one I have here called as species typical duet song type sung by the hill slope birds and the second I have termed as repertoire singing type-2 (see Table fourteen,, chapter Four) sung by the plains members but only in the season they interbred. The second type-2 is comparable to the repertoire song of the magpie robin and like I remarked there this kind of repertoire singing seems to be associated with intergrading species or members. My description and assignation of this song type-2 is further derived from some important observations. It may be noted that the intermittent interbreeding is a one way process. It is always the plains member that inducts a hill member for breeding and not the other way

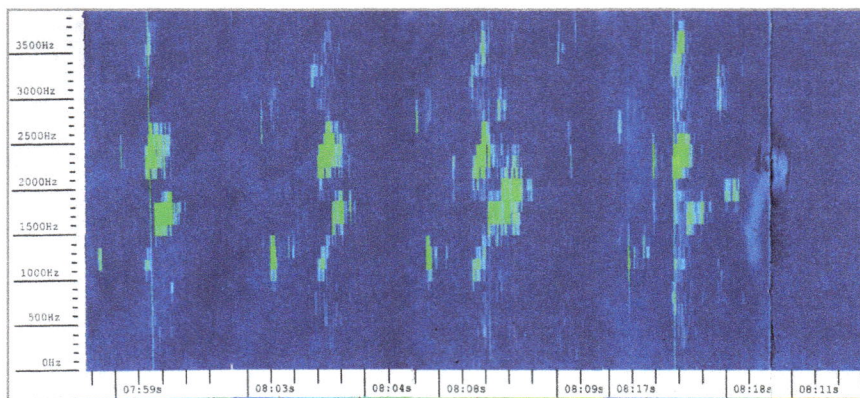

Figure 8: A line from the tree pie (parvula male) breeding song corresponding to the notes "kukramyaku, kukrakyaku"

even though both of them translocate and exchange territories at this event. Since the two populations are faced with two different handicaps in their habitats (as I outlined above) they seek to circumvent them in this manner. Thus cyclically once in every three to four seasons the breeding pair from the plains is composed of two members from a clutch raised on two different ground types. And it is they who produce this repertoire song.

It is my conclusion that the type of repertoire singing produced by the magpie robin and the parvula is typical of intergrading populations of same species (see Table Five in section on Tailor bird for a comparison). This song is made up of rearranged component notes of the species typical and the song routine reflects what I have termed elsewhere as "here and there song" The female parvula produces her song solo season initially in August and then again at the end of season November -December. The male produces during peak breeding season between hatching and fledging. The song notes are sex differentiated. In Table Four below I have attempted a comparison of the vocalizations produced by three members belonging to this same family and that share the territory.

The hooded oriole female one note covered more than three months but the male song spanned only six days of the peak breeding season. The pallida's duet songs spread across five months. The parvula female's song also is spread over five months with a gap in between. But the parvula male produced only peak singing that covered a fortnight. The songs of these birds fall in the category what Marten *et al.*, labels as "verse song" (Monographs, 2011, p. 88). Verse songs are more elaborate and are made up of varied notes. They also display time breaks between a set of notes, which makes them into lines or stanzas when the lines are repeated. Not all the lines may be repeated. In the case of the songs under analysis these time breaks are of two kinds, across time of day and over different days; so that the song begun on one day stretches over the next for five days, (I suggest that these songs should be seen as one continuous song stretching across many days, rather than break up into songs of each day or every day time period) The notes produced in each day are not the same as the before. Among the tree pies there is male –female duet singing most of the time and their song stretches over several months. There is more frequency of repetition of all the lines. The hooded oriole male's song lines were never repeated.

Table 4: Comparison of vocalizations of the Hooded oriole, and Tree pies

Hooded Oriole (BM)	Hillslope Tree pie (Pallida)	PlainsTree pie (Parvula)
kraunck (f)	krun (im)	krun (f)
krau (m)	kukrakmya (im)	kukrakya? kukrakya? (f)
krayaru... krakra kukra	kukrakukra	kukakru? (m)
kukru (m)	kyaka (m)	kukrakmya kruk (m)
kra$_{ss}$ukra$_{ss}$u kukra	kukra kukra myaku (f)	` kukra ` kukra (m)
kukru (m)	kruk (m/f)	kukra myaku? kukra
krau krau kruwu kruwu	kragudukragudu (f)/	kmyaku? (m)
ukru ukru(m)	Krang (m)	kukryaru?kukramyaku?
		kukrakryaku?
		kruk(m)
		kukramyahu kukramyahu(f)
		kukramyaku kukramyaku
		kruk(f)

Legend im= immature; m/f= male/female; bm= breeding male;

I suggest that the three bird species to a great extent of the notes are singing the same notes but in three different dialects (suggesting cladogenisis), *viz.*,compare notes krayaru , kragadu & kukryaru? Or kraunck, krang & kruk. Other notes are produced in different orders in terms of placement of lines (or stanzas) and in terms of note break up, but not in terms of combination. Only hooded oriole produces an additional and distinctive note, *i.e.*, kruwu kruwu. I suggest this note sets them as distinctive and independent species from the tree pies. It is possible molecular analysis will corroborate this genetic difference from the tree pies. Other differences are not due to introduction of new extra notes, but differences arising out of species specific meanings, *viz.*, krau & ma denoting females of the respective species; ukru & kukru denoting modes of nesting in open canopy and hidden under canopy, respectively.The parvula tree pies are an intergrading species and their song notes follows that of the pallida albeit wherein the original notes are broken up and recombined differently(repertoire singing type 1; see Table Five section on tailor bird).There is only one note which is not found in the pallida song and seems to belong to the oriole's *i.e.*, kukryaru.

Below we shall review the situation of the tailor bird who I have claimed produces song type mismatchingas well as repertoire singer in a community of bird which is dominated by genetically diverse species.

Tailor Bird

The tailor bird (*Orthotomus sutorius*) is another popular bird whose nesting habit is a matter of popular mythology. Even though mistakenly thought to be tailor made, the leaf nest indeed has an intriguing method. A breeding bird produces an oral secretion (equivalent of saliva) that when dry solidifies to fluff (looks like white candy floss; see section on sunbirds for more on this). The leaf nests are made by gumming together with this saliva. The fluff also lines the insides. When the chicks hatch out their increasing size gradually releases the loosely gummed leaves to make a cup-like open nest. Now the bird's talent is not limited to nesting only it has great singing talents as well. The typical song is composed of the syllables "tuwituwitu" repeated at regular intervals. But they can produce clever variations of this typical song such as "tuwituwi, wituwitu" or "tutituti, wutiwuti" at will when provoked. All these vocalizations are delivered sharply and in such a manner that the bird's difference from others in its territory is left in no doubt. This delivery can be interpreted as song type mismatching and as signaling a truce and neutrality in the territory. It marks out the bird as having a niche of its own which does not infringe, they build a nest like no other bird can, and forage on ant eggs instead of grubs (favored by flycatchers and others) and worms (favoured by ioras). Tailor birds are in fact very versatile they can inhabit grass lands as well as edges of forest with plenty trees. They forage in the lower levels as well as the higher levels of trees if need be.

Their song versatility is proved further by their ability to learn new songs and to produce repertoire singing. A tailor bird can pick up songs throughout its life span from any kind of social tutor a bird, a radio or a child learning her lessons and reproduce them at will anywhere in its territory. But they possess this song expertise only at the cost of subordinating it to the dictate of other birds in the community. Resident birds which are temporarirly absent during breeding time from a foraging or roosting location induct the tailor bird to play the role of stand in for them and make it to imitate their song. It is they who transform the little tailor into a repertoire singer who learns every song in the territory. It is possible that this sort of repertoire singing of the tailor bird provides camouflage in addition to reserving their territory during their absence from any possible interloper. The song learning of the bird indicates the existence of reciprocal structures of dominance in a community of birds, the tailor bird can imitate fellow members to their benefit only. Thus the tailor bird is a key in the inter species dynamics of a forest edge habitat.

One of the most noteworthy vocalization of this dynamics is its alarm calls. The bird produces the alarm calls not when it is threatened but when another species member is. For instance a young just fledged shikra male will take cover under the alarm calls of a tailor bird in his territory in order to distract an older but still juvenile predator member who has similar (natal) claims over the territory, such as a young yet unpaired brahminy kite. He himself remains quiet and hides under a leafy bough of a tree. The kite youngster emits a piercing mobbing call as it continuously circles the hiding spot overhead in the sky. Perhaps the loud high pitch intensity alarm calls distract the predator from the actual location of his real

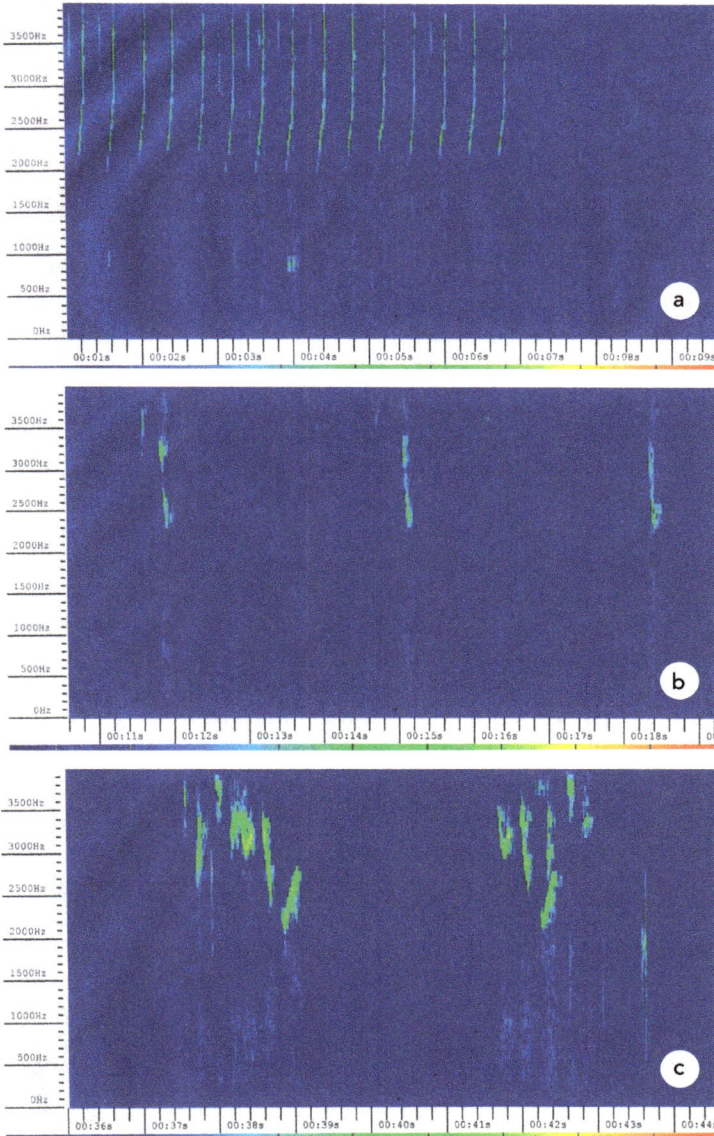

Figures 9a, b & c: Two different vocalizations (a typical note "tuwi" and its reverse "witchun", respectively) of a tailor bird. 9c displays imitations of other birds such as bulbul & leaf bird produced by the same member

prey. The cost of the signal for the tailor bird is far less than the real prey because the predator is not really interested in it and moreover because she is female (see below). Females among bird species have a prerogative because they seldom lead an interspecies confrontation. For instance a shikra female will not hide under a canopy like the male does when mobbed by a predatorial kite; instead she proceeds to sit on top of tree canopy clearly in the open on such occasion. I suggest that the tailor bird vocalization is another example for kin group selection and reflect reciprocal structures of dominance.

But by my account of the vocalization capacity of the tailor bird given above little will have any one guessed that this talented nest building and singing bird is a female. And that she is the only one bird in her flock that can sing. In fact among the birds surveyed in this book only the tailor birds have a lead singer who is female. Even I was taken by surprise when I discovered that the singing tailor bird I was keeping an eye on was a female. This discovery hit upon me when I found her building nests which on inspection turned out to have a single grayish rotten egg. It couldn't be that a male bird can lay even a nest of bad egg?

Actually tailor birds are not sexually distinguishable morphologically. According to the field guides (vol 8, Ali & Ripley, 2001; Grimmett & Inkskipp, 2005; Grewal, 2005) only the shorter tail in the breeding female distinguishes her from the male. But the phenomenon of the nests with bad eggs indicated that the singer among tailor birds is a female and she is the only bird in her flock that produces any vocalization. Tailor birds are usually found in pairs in their territory season initially, *i.e.,* April- May. By July one is more likely to find them in a flock composed of two adults and not less than three chicks. Neither the pair nor the adults with chicks produce any vocalizations of any sort when they move about foraging. Singing is the prerogative of only one bird at any time of the year. Early in April the singing bird may be found participating in paired foraging but she does not produce any song then. She seems to split from her flock once she gains capacity to produce song. Thus songs are produced solo and when alone. In fact the singer occupies a portion of the territory apart from her flock and sticks to her part.

The most intriguing feature about this repertoire singing female tailor bird is that she goes about building several crappy leaf nests in several places almost all over the territory on different trees during the peak breeding season May-June. During this time she also molts seemingly an untimely one. Molting is a period in any bird's life. It is accompanied by hormonal and other bodily changes. I have

Pictures 23a & b: Tailor bird breeding female & Singing female in molt

seen molting birds show visible changes in mood, sulk or turn grousy (cuckoos, prinia), sing irritatedly (tailor bird) and become quarantined (orioles) to a part of the grove plentiful in certain types of foods such as grubs rich in calcium I believe. Our tailor bird does not overcome molting until she has finished with three batches of bad egg. I investigated the first nest closely and found it with a single grayish looking tiny ball, which I assume was a bad egg. After this she seems to have found tranquility but by now she has lost her species typical song and become a repertoire singer who imitates sounds in her environment.

Let me review what I have been saying about the tailor bird: In a territory that is inhabited by tailor birds there is only one singer among them. She always sings solo and seems to be outcasted from her flock. And that she lays bad eggs. In ornithological literature bad egg-laying is linked with hybridity. That would mean our tailor bird is a cross bred. Another way of ascertaining hybridity of a bird is by looking for resemblances with the cross breeding parent species. Our repertoire singing tailor bird does not have any morphological difference from other tailor birds, and most importantly it is she who produces the famed species typical song. In addition to this the singing-bad egg laying member is to be found in every season's flock with a clock work regularity. There is nothing accidental about her presence as hybridity would suggest. A singing tailor bird is unique to the clutch she is born in, the next season's clutch produces it's own new singer. I have positive evidence for this since at the period when last year's season ends and overlaps with the beginning of the next, one can hear both the birds in two contesting locations.

My suggestion is that our tailor bird is not a hybrid but that the sexual ambivalence about the bird is owing to her prolific song learning urge she finds herself in a state what in a male is known as neutered. Our singing tailor bird is a female neuter bird. In this hanging state she cannot completely be inducted into pair breeding and remains a third factor in a broad coalition. The singing tailor bird demonstrates our hypothesis that song capacity is inversely related to reproductive capacity whatever be the sex of the bird. She is a good example of what would happen to a female lead singer if she continues to produce song all her life: she would end up lay inviable eggs. She becomes sexually inviable. Considering her role in the breeding coalition there could be a possible explanation for her nests of bad egg, they are meant to divert ants and other insects away from the real ones in a real nest in the same territory. The case of the singing female tailor bird parallels that of the singing males among iora and magpie robin. She loses her prerogative as the breeding female owing to her devotion to song. But this does not deter her like her male counterparts among other song birds she still plays her role to the hilt as the third factor in a breeding junta contributing equally to the propagation of her species. Like I said in the beginning of this book Nature is full of cunning and camouflage that it can leave the human baffled and in a daze. In Nature each has his/her role that is important to the well being of the life world as a whole.

Below I present a table comparing the different types of repertoire singing observed among birds of the Bhadra. By the comparison found in Table Five I have sought to distinguish between two kinds of repertoire singing. Type 1 refers to song production that is made up entirely of the same notes/ syllables of the species typical song but the notes are broken up and the constituent syllables are reorganized in all/several possible permutations and combinations. Type 2 refers to production of imitations of songs of other birds in the territory. The tailor bird produces both types. Type 1 repertoire singing is almost always preceded by singing

Table 5: Repertoire singing displayed by four passerines compared

Repertoire singing type 1		Repertoire singing type 2	
Magpie robin	Treepie (D. v.parvula)	Tailor bird	R.W. Bulbul
tweop (off season)	krun(f)	tuwituwitu ⎤	Tweet tweetrio
song type 1	kukryaka? kukryaka?(f)	tuwituwi,wituwitu⎟	kra pip
twe,twetavi,twe	kukakru?(m)	tutituti, wutiwuti ⎬	twitwittwit
twe,twetavi, twetavi,	kukrakmya kruk(m)	**song type 1** ⎟	
twe	`kukra`kukra(m)	⎦	
song type 2 (SM tree	Kukramyaku,kukra	tchweop tchweop ⎤	
top repertoire)	kmyaku? (m)	twon twon ⎟	
twetavi (3times)twe	kukryaru?kukramyaku?	twittwittwit ⎟	
twe(3times)twe tavi	kukrakryaku? (m)	tweettweeter ⎬	
(3times)twe tweeeee	kruk(m)	**song type 2** ⎟	
eetwe tavitwee twe	kukramyahu	**(mimicry)** ⎦	
twee, e,ee,twe,ee,	kukramyahu(f)		
ee,twe,eee, eee,	kukramyaku		
twetavi(3 times)	kukramyaku(f)		
ee, tavi (3 times)	kruk(f)		
twetavi 3times)			
twee, whistle, ee,			
whistle,....... So on			
Song type 3 (SM			
IInd season)			
twetavi,twe, weeee,			
ee .... twe, ee, ee,			
twe, ee, ee, ee, twee,			
tweee, twee, tweee,			
tweee...tweee...			
twe,ee.tweee....			
eee...ee, ee, ee, ee,			
ee, twe, eee,twee....			
twee ... ee....ee ,,,,			
so on			
twe tavi, we, ee, ee,			
twee, ee, ee, tavi,			
wee, ee, ee ,			
twee, ee, wee, twee,			
wee, ee, ee,			

Note: I have presented only first few lines of the Magpie robin's three different song types. The third song type for instance extends over several weeks and may not be presented here, while song type 2 is sung for half an hour at a stretch. These song types are sung in different seasons and at different stages of the breeding. Similarly the tree pies song of the species typical song. But in the table below we may note that the parvula does not seem to produce the species typical song that is found among its sibling pallida (see Table 4). But rather it produces type 1 singing from day one of song production and its song is entirely made up of broken up notes of the pallida song. Only one note is borrowed from the hooded oriole's song which shares territory with the tree pie. It is also possible to make another distinction the tailor bird's type 1 singing is delivered sharply what I have noted as mismatching in the above section on the bird, a phenomenon not found among the other two birds compared in the table. The comparison in the table suggests some generalizations on repertoire singing. It is likely that species that intergrade into other's territory produce this kind of song matching. Secondly there are different kinds of intergrading as indicated by the different kinds of repertoire singing.

Jungle and Scimitar Babblers

The babblers are a family of most gregarious and sociable birds in the Indian sub continent. Ten species are found in the Bhadra region, in little parties of noisy vociferous birds. Three of them (jungle, a subspecies of jungle and scimitar babblers), are permanent residents; the rest are migrating species moving from glade to glade en route like the thrush. The scimitar babbler (Pomatorhinus horsefieldi) may be distinguished by its sonorous vocalizations and its preference to move about singly. It is also a resident in the territory heard all through the year.

Babblers as a family are remarkably socially intelligent. Not only do they breed and forage in a sizably large cohesive flock, a territory (as wide as 230 acres) may be demarked and shared amicably by the various species all shunned by other birds for their voracious and noisy feeding habit. The resident larger sized jungle babbler will hustle a curious smaller sized tawny bellied around its territory. A

Picture 24: Tawny bellied Babbler

visiting migrant sister species such as dark fronted babbler can be given a guided tour. A social visit to one's location elicits reciprocity and the babblers are never too shy of such social rituals. But there is no question of rivalry between these ten species of babblers. The leading male who is usually younger than the breeding members and who plays a surrogating role in the breeding coalition will puff out his streaked neck (bulla), shake his chin reprovingly and stare down with his typical ferocious eyes perched from the lower rungs of tall jungle trees, but never stop a curious visitation (this may sound blaze for a scientific book but in truth this is a typical pose of the male jungle babble and is testified by naturalists like Amotz Zahavi). Among the resident species even mutual exchange of breeding territory may be practiced seasonally.

Vocalizations make an important means of group cohesion even when a species don't sing. Most of the babblers are strongly flocking species and breed in sizeable flocks. The size of the flock varies among the species depending on the size of the bird, larger babblers are found in smaller groups smaller sized like the tawny babblers are found in larger number flocks. It appears that a babbler breeding male is a polygamist so that there is only one breeding male to a whole flock of females with one leading female breeder (it is likely they are all neutered members excepting one reproductive female). Such a breeding flock is adopted by a younger male usually born of the same clutch as the breeding female. He will play the leading (but non-mating) male role in that flock by taking on a surrogating parental role and at the end of a breeding season he is rewarded with a mate so as to enable him to start a flock of his own. He plays a prominent role in safeguarding breeding territory and in leading the foraging party by producing strong visual bodily displays as well as vocalization. The leading female can join in the vocalization but strictly their roles are demarcated; the chest puffing and chin wagging display is unique to the surrogating male. The breeding male does not produce any display. For instance, among the jungle babblers such a young adult male generally has a leading role in taking the flock from place to place both during breeding and after fledging. His distinctive, brief, abrupt, and authoritative "kraack" marks him out as male and as leading. When he calls out "kraack" once or twice like this the rest of the family will hop after him, with the lead female answering in a continuous "krakrakrakrak" as if to mean "we *etc.* follow suit". There occurs a change of leadership with the leading female taking over when his chosen female in the newly fledged brood comes of age. At this time they separate from the flock and engage in courtsip invovling preening their feathers while perched close together. The leading female will take over her flock producing vocalizations, albeit on a different note from her earlier vocalizations but including a part of the male's (kraackkak or kraackakakak) as if she was playing both the roles. This flock will rejoin the young mating pair once the mating is over. Other species of babblers in the territory have the same mode of group communication even though vocalization may be produced on a different scale depending on size of the species. The jungle babbler breeding season falls from March to November and the common (likely a subspecies of jungle) babbler's from June to October.

Picture 25: A young surrogating common babbler male with his newly acquired female

Among the scimitar babblers there is a sex reversal of displays and the surrogating role. Unlike the other babblers they form loosely cohesive groups with likely a breeding pair and a younger pre-breeding female. This female in her pre breeding comes out of the same clutch as the breeding male but is younger and allopatrically diferentially bred (meaning at different locations within the same territory). Vocalizations of two members can be heard distributed through out the year, one that of the breeding male and the other the non breeding younger female. It is a bit difficult to say accurately the sex assignation of the second bird because the scimitar (as also the other babbler species) is not sex morphed, but it is a good conjecture that the younger is a female in her pre pairing stage and awaits her turn. She produces a great deal of bodily displays in addition to the song; such as rolling swinging *etc.* which I find common among young adult females of other passerine species, iora, oriole, leaf bird, minivet, white browed bul bul (pre breeding female displays). Also this can explain why the breeding female does not produce any vocalization. It is a fair conjecture that once this pre breeding young adult finds a mate and enters breeding she ceases to produce any displays. A case that parallels the surrogating-display phenomenon among other babbler species such as jungle and common I described above. Vocalizations form an important means of cohesion in this coalition also. Different 'songs' or rather sonorous vocalizations correlating to their status in the breeding junta and season is produced. The songs are so cacophonous and kept up almost through the year, any one passing through a jungle here abouts is reminded of a restless wandering soul laughing as if in some deep purgatory. They inhabit thick secluded forests areas and patrol their territory noisily morning and evening from one end to the other, only their cacophonous laughter telling of their movements.

In the beginning of monsoon mid May - early June (peak breeding season) one can hear the sonorus throaty rolling "kokoko kok, kok kokoko" songs of the breeding male probably similar to a hen after laying her eggs perhaps signifying success in breeding. Gradually this song becomes more low-keyed "kokokok" that ends on a questioning note. The number of "ko" varies from one breeding season

Picture 26: Scimitar Babbler

to another perhaps signifying the breeding age of the bird. On the other hand his deputy the non breeding female produces a less talented scolding "kokosk"(that sounds like a battalion practicing marchpast) rest of the year, *i.e.,* from September to May. This vocalization is produced while migrating from one end of the territory to another several times through out the day, likely territory ranging between the actual nesting site and roosting/foraging site. She quiets up once a breeding male takes over who also duplicates her movements. Briefly in the beginning of monsoon one can hear both of them together as if one was taking over from the other patrolling the territory. This makes it clear that there are indeed two different birds. Not only is the song of the latter invariant but it is also produced for a greater part of the year (off season). It is possible that the surrogating younger bird contributes to allopatric foraging and maximizes the niche.

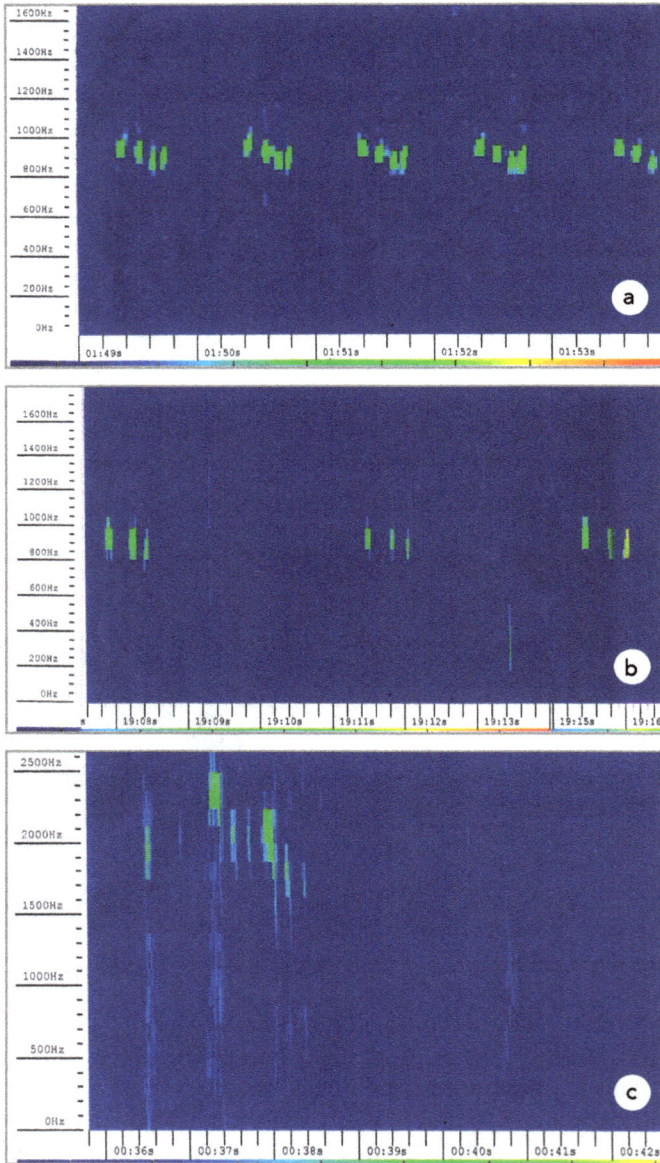

Figures 10a, b & c: Two different songs from the Scimitar's breeding repertoire (ko,kokroko & kok,koko, roughly corresponding to mid and late season); 10c shows vocalization produced by the jungle babbler surrogating male

There is no evidence of song learning from the immediate surroundings among the scimitar babblers. Still there is a chance that these songs were picked up from the environment by a generation in the past and have become genetically coded. Perhaps one will find more comparisons and similarity between scimitars and other babblers such as laughing thrush but these are not to be found in the Bhadra area. The babblers as a species seem to use vocalizations for group cohesion *i.e.*, for keeping together in their wide territory. The noisy vocalizations also indicate their prerogative as a group to move freely in a locality. They are rather like a gang of teenagers stalking a park. Scimitar vocalizations also seem to work in this direction, they are communications between member birds of the same coalition and also seem to mark their prerogative to be there and directed to members outside it.

It may be summarized on the bases of field observation presented on the babblers in this section that displays are to be viewed more as aspects of breeding systems than merely a mating pair. Further it is not mandatory that the breeding male produce the displays. Among the babblers we find that a younger male (female among scimitars) who has not achieved full breeding capacity and has been superceded by an elder mature male (female among scimitar) at that point in the breeding coalition produces the displays. His or her displays are a dimension of the surrogating role he/she takes on in his/her pre breeding stage rather than breeding. This I suggest challenges the currently held theory that displays are for sexual selection. It is more likely that displays are for kin group selection albeit on a reciprocal basis where both the parties benefit from the coalition. In the case of the babblers the younger surrogating male's displays are produced in the situation of an older mature male establishing a prerogative over the possible females in his birth flock. In this circumstances the best compromise he can achieve is to direct his display at herding the reproducing flock in order to acquire a female. This arrangement is similar to that found among Tickell's blue flycatcher (see section above). It may be remarked that the babbler displays are role specific rather than sexual. Role distribution among a flock enables each and every member of a species to maximize the niche and it makes more sense to form a coalition of interests than to simply carry on individualistic contests for the best mate.

There is one more very important field observation that concludes my generalization on displays as kin group selection and babbler bulla wagging display as role specific. It must be noted that the actual breeding bird does not produce any vocalization nor does he produe the bulla wagging. Why is this? The reproductive feeding behavior of the babblers (and perhaps other birds also such as grey breasted prinia) points to the fact that the bulla (part of the gullet) of the bird is also used to carry stones to grind food for the purpose of feeding the incubating members (The bulla can be also used to transport wet clay soil for building nests. Such nests are built at the base of a tree with outgrowing roots). It appears that younger birds that produce vocalizations have not yet learnt the trick of keeping the stones to grind food so much vital for the reproduction as an adult member can do (see tree pies for a parallel). No doubt of course the adult member has lost his capacity to vocalize by this. But then the younger bird therefore can effectively play the surrogating role to safeguard the territory and lead the flock. Thus displays are governed by a biological economy that capitalizes on group selection.

Bulbuls

Bulbuls are next to babblers in capacity for vocalizations that are kept up through out the year. These sounds are pleasing and musical but not exactly the elaborate compositions of the iora, magpie robin or Tickell's. These gregarious birds are found in diverse species everywhere in the Indian countryside, in gardens, clearings and fields. The vocalization of the red whiskered displays aggressive sociability what in the literature on song has been called as mobbing behavior. Their species typical vocalizations can turn into mobbing calls of varying intensity depending on the type of intrusion into the territory. A traveler walking in a forest clearing can be accosted by a pair of these birds and lead by the nose as it were to the exit by a pair of aggressive vocal red whiskered members, while the rest of the birds in the territory go into hiding and fall silent.

In the south Indian forest habitat in winter there are sighted three species of bulbuls. The red whiskered (*Pycnonotus jocosus*) and red vented (*Pycnonotus cafer*) are often found in small parties during off season. Only the white browed (Pycnonotus luteolus) are found in pairs all the time. Only a discerning observer can distinguish them by their vocalizations or plumage. All of the three are various shades of dun. Vocalizations all are made of "tweet" and "r" roll sounds of different shades and seems to be variations of the same song. They build similar looking cup nests but each at different levels of foliage tree top, middle rung or undergrowth and on different trees or shrubs. The nests differ in the use of materials, creepers, thorny shrubs or grasses.

Picture 27: Gang of Red whiskered Bulbuls on a bamboo shrub

Even though to casual observation it appears that these bulbuls occupy overlapping territories with similar habitat type - wooded areas - this appearance is highly deceptive. They in fact occupy distinctive territories distinguished by specific ecological conditions. These territories are strongly fixed and marked and exist in a structured relation between them. The birds strictly do not intergrade and show minimal secondary contact. Let me explain; the red whiskered occupy hill top forest clearings, the white browed the scrub slopes of these clearings and the red vented occupy the wooded areas on the plains. In addition these territories are situated in fixed directions to each other, the white browed territory is to the south east of the red whiskered and red vented to the northwest. The song capacities of these bulbuls and the inter species dynamics of song I suggest reflects this fixed territorial mapping between them, especially red whiskered's versatility in song learning which can be grouped as repertoirie singing of intergrading species I noted earlier.

A comparison of the songs of the three species in this region indicates that the red whiskered has the simplest typical vocalization but predominates by its capacity for song learning. The normal vocalization of the red whiskered is "tweet tweeter" the simplest; that of the red vented is a more elaborate rolling "t" variation "twittwittwittu" and that of the white browed

Picture 28: Red Whiskered Bulbul

bulbul is the most elaborate variation of them a series of rolls "twitertwitertwiter". The red whiskered with its simpler songs have developed ability to learn new songs, versatility that the other two do not show. In my region the bird has learnt to produce expressive vocalizations that mimic school children reading their lesson in the nearby area that goes on like this "tweet, tweeter? Tweet, tweeter." The songs produced by individual members that share the same territory show remarkable variations such as "tweeterio, tweet, tweeterio" as against "tweet, tweeter? Tweet, tweeter." Most of all they can produce good imitations of both their sister species. This perhaps involves a semiotics of communication in itself and should be recognized as a kind song type matching. But when they do adopt the vocalizations of other bulbuls they do not produce their own species typical tweeter. In fact the bulbul species – al the three species- do not produce any vocalizations even their

Picture 29: White Browed bulbul

species typical until they have achieved reproductive success. Young fledglings are very silent until this time. It is the adult bird that produces the vocalization. It is possible that this explains why the red whiskered bulbul that learns the song of other species fails to produce any species typical sounds.

Table 6: Comparison of song capacity and song types between species belonging to same family

Bulbul	Leafbird
Red whiskered	**Cochinchinensis** wee ´ wee ´
1)tweet, tweeter	wee wee chipchip
2)tweeterio, tweet, tweeterio	wee wee chupchup
3) tweet, tweeter? Tweet, tweeter	chipchipchipchip pip
4)tweeter, tweet	
5) twittwittwit	**Insularis leafbird**: wee wee wee wee
	twwieue druwitu druwitu
Red vented twittwittwittu	wee wee wee wee pip pip
	pippip pippippip
White browed: twitertwitertwiter	

Note: 1, 2, 3, 4 &5 vocaliztions are records of five different members of red whiskered. Among the other two bulbuls vocalizations are invariant across members.Cochinchinensis and insularis vocalizations data all belong to only one member.

It is clear that among the three bulbul species song capacity seems to be distributed differently. If the red whiskered have more song learning capacity and produce more volume of sounds, the other two produce more elaborate species typical vocalizations in comparison. It is difficult to make out if there is sex differentiation in song capacity because there is no distinct sex differencing plumage in all the three. Reproductive participation seems to be decided by prerogative of age rather than sex; an older pair in a flock has the privilege of raising the chicks even that of a younger pair. This is similar among babblers also, males among them can have more than one female attached and a younger female is always favoured. Here we find another type of breeding coalition between cross generational pairs. Ultimately propagation of species is a co-operative affair rather than pair breeding.

In table Six I have attempted a comparison between the song capacity and song types between birds within the family clade of bulbuls in order to draw some

Pictures 30a & b: Red vented bulbul subspecies members in the region

generalizations. It is possible to generalize that the song type matching that is found among them, *i.e.,* production of a more elaborate song, is typical between member species of a family clade. Secondly this suggests cladogenesis or phylogenetic origin. There seems to be a comparison with the interspecies dynamics among leaf bird species, which as we saw elicited generalizations about the speciation relation and challenges to the use of ecological speciation terminology. It seems even though allopatry is a necessary condition for speciation, it is not the primary or even suffecient for new species to come into existence. Allopatry makes available diversity of niche and perhaps therefore catalyzed speciation. Speciation phenomenon must have deep causes and guided by inner protocol.

It is well documented (Grewal, Bikram, 2000) that the bulbuls diversified from the North eastern Himalayas like the leaf bird. The three species appear to have diversified through different routes: the red whiskered must have originated from a species of Himalayan bulbuls accompanied by gradual expansion of range. The red vented and white browed speciated facilitated by drastic or sudden allopatric changes, very likely migration, to their different regions. Their present range is a secondarily extended range over the Indian peninsula. The structured (similar to colony formation in fixed geometry among bacteria or other microorganism) territorial distribution among the bulbuls in this region points to correlation between the distribution of local populations of a family cluster and the same family cluster's global distribution in its range. Species family tend to replay their global or macro territorial relations at the micro distribution levels, apportioning local territorial differences in much the same way as exists at the level of their macro distribution. It is clear that even though the ranges of the three bulbul species seem to tbe overlapping they are not overlapping. They remain diasporic in their expanded range. The structured territorial distribution is also evident among the subspecies populations. Perhaps this indicates them to be intermediary forms and the result of disruptive adaptation for species cohesion. For instance among the red vented bulbuls in the region at least two subspecies may be recognized that fit the descriptions and distributions presented in Ali & Ripley's (vol6, 2001, p. 86) Pyconotus cafer cafer and P.c.humayuni the former distinguiished by darker plumage and more extensive black hood than the latter. These two occupy locations within the territory in a structured way, the former occupying higher slopes and the latter the foothills and remain non intergrading. They also display differences in song capacity.

Purple and Purple Rumped Sunbirds

The sunbirds are a large family of small sized birds which resemble a great deal the hummingbird (of Trochiliode family, related to swift of Apodiformes) but are not assigned the same status. Not only most of them have shiny shades of plumage comparable to that of hummingbirds, they display hummingbird like feeding behavior, *i.e.,* they suck from flowers while hovering by continuously beating the wings. This method I observe is to avoid dislodging the flowers from its stem and calyx while sucking the nectar. Because a flower can replenish the nectar again and again until they are mature to be pollinated. The sunbirds adopt the hovering method in order to maximize their food resources. Not only this, a hovering bird can delay the pollination by minimizing the contact and this is good for the flower because fertilization at a reproductively inviable age is avoided by this. The sunbirds produce song almost throughout the year and make up a large portion of the song heard in a South Indian forest habitat. The males exhibit what has been labeled as "endless singing" a few simple notes repeated all through the day while foraging on tall tree tops. The voice is clear sounding with a tinny musical timbre and pleasing to the ear. The females on the other hand produce an elaborate verse composition running into around seven days during peak breeding. The syllables are also sex differentiated (see table below). There are five species of resident sunbirds in the region.

The sunbirds share a similar reproductive system among them which involves only the breeding pair. But in a territory one is bound to find pairs of intermediary forms exhibiting differences in beak and body size. These differences are accompanied by differences in the flowers frequented for foraging. Strictly different parts of the same territory are used for nesting by such fellow members of the same species as well as the different species. Going by Ali's inventory (2002) it is possible that the purple sunbird intermediary (picture 3 in the grid below) and the crimson sunbird intermediary (picture 6 in the grid) are the same as the subspecies labeled as loten's sunbird (*Nectarinia lotenia*) and small sunbird (*Nectarinia minima*). But it must be noted that Ali's description does not fit the intermediaries I recorded in any accurate manner. It is my suggestion that those identified as subspecies (by Ali and others) are actually not so but merely intermediaries, that owe their existence to the allopatrically differentiated reproductive praxis of the parent species; And that this explains their persistence across the entire Indian range perhaps even outside it. It appears that these intermediaries are born in the same clutch as the main breeding pair but later allopatricaly fledged in different locations of the same territory and very likely are made reproductively inviable by it even if paired off. The role of these intermediaries is not clear but it is possible that it is to safeguard the species unique beak structure. The sunbird optimum beak structure is so exhorbitant that it makes the member opting for it inviable. So some members always choose to have a shorter beak and retain reproductive capacity. It appears that the optimum beak

Pictures 31a, b & c: Purple sunbird and its intermediary forms (a, b & c). Crimson sunbird and its intermediary forms (d, e & f)

structure is the mark of sunbird species ability to adapt to the diversity of micro niche and therefore good to keep. So a compromise is reached by the distribution of roles at the group level; some members retain the extravagant beak as a species features and the other not opting for it undertake to breed. In the following I present observations on the purple and purple rumped (*Nectarinia asiatica* and *N. zeylonica*) sunbirds. These two species display different song patterns and have different song routines. There appears to be a complementary distribution of song routines between the members of the two species.

Among the purple sunbirds the breeding female inaugurates the season's song routines in January peak breeding, with the male contributing a small quantity in this period. But the season ends with the male's song routines and continues with his one note vocalization up till August. Among the purple rumped there is

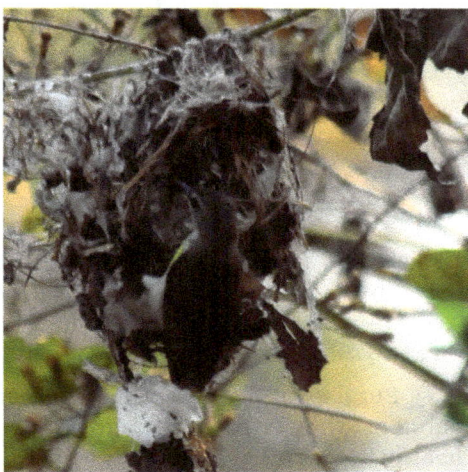

Pictures 32a, b, c & d: Purple Rumped Sunbird male, female & nests

an inversion of this routine, with the breeding female's elaborate composition in January bringing to an end the breeding male's song routines extending over November -December. This inversion has a correlation in the song pattern and the notes that make up the song. As can be seen in the table below among both species the breeding female produces a more quantity of song in a short period of time while the males produce less quantity distributed over a longer period.

The sunbirds have an interesting nesting arrangement. Their nest is actually made up of leaves gummed together by a secretion got out of the mouth during breeding. Juice of the belladonna flowers or poison ivy or nectarine or lantana is used to produce the nesting saliva. The juice is had in large quantities and then clean water is consumed. This makes the bird to regurgitate the saliva in the form of a frothy secretion which when dry forms a cottony fluff that lines the leaf nest. (Tailor birds and minivets consume certain types of worms to produce the nesting saliva.) These leaf nests are situated on trees and serve as incubator for the eggs. Once the eggs hatch the fledglings are transferred to hanging nests made of plant fibres and the fledging takes place in the second nest.

Picture 33: Purple Sunbird at belladonna flowers

The hanging nests are usually built from nests of other birds such as pale prinia and flower pecker. But the purple rumped sunbird is unique in that it builds its own nests with much dexterity while other sunbirds especially lotens and purple sunbirds steal the hanging nests of others. This I gather is because the sun birds are not equipped with suitable beak structure to weave the fibres tightly even though they may adapt skillfully discarded nests of other birds. Such nests are lined and made stronger, the hatch over the opening may be strengthened, a ledge may be added for the parent birds to use as foot hold while feeding the chick. The hanging nests are made by using the saliva to loosely matt together grasses and thin creepers rather than weaving them. The fluff may be used decoratively to camouflage the nest from predators. Both the male as well female in the breeding pair contribute to the nest building, even though the male seems to have the role of helper.

But the strangest thing is that the female sunbirds are not fledged in such a manner by transferring to hanging nests; only the male has this privilege. In appears that the females fledge rapidly and simply fly out of their incubator nest. It is a fair guess this sex differentiated nesting habit also causes the differences in plumage pattern among the sexes as well as differences in song capacity. The song routine of

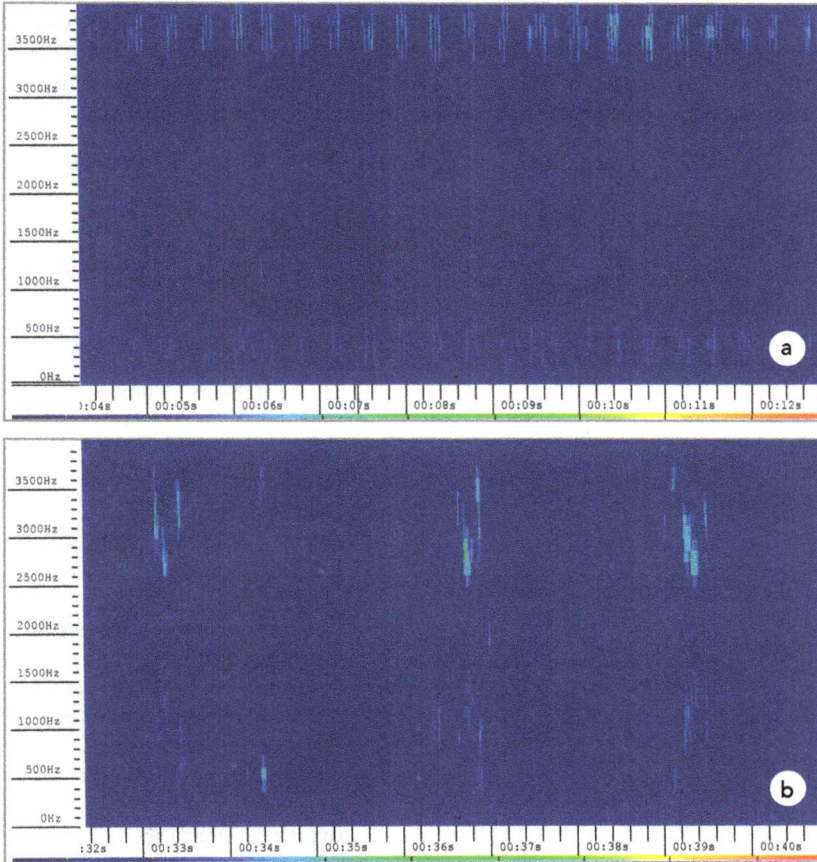

Figures 11a & b: Purple rumped sunbird (male, che, ple notes) and purple sunbird (female, tche, chewere notes) songs

the males in both the species also supports my hypothesis about repertoire singing that involves the break up of notes to their components produced by members that intergrade. We have already looked at this type of song in the magpie surrogate, tree pie and tailor birds. Among all these birds allopatrically differentiated nesting is practiced from one season to the next so that each generation of members differ among them. I suggest the dual nesting habit of the sunbird should be viewed as a kind of allopatric differentiation and that this manifest in the song type. In Table Seven below I have presented the vocalization data for the two sunbirds. The data is limited to only certain portions of the song which stretches over the entire season peak for about three months. Analysis of the purple sunbirds female's song routines- she has at least three - suggests that she begins the season in January with an identity song. This song is delivered from tree canopy. The second version communicates the mode of building nest and the third records the proceedings of the season, tracing the various stages of development of the eggs and the hatchlings. These two songs are delivered from middle rungs of trees its verses spread across the months of March to May. During April the female bird descends from tree canopies flitting from belladonna shrubs to a clean water source accompanied by the male. Her song comes to an end in May when the male takes over with his simpler song. If you notice his song reverses the notes of the purple rumped sunbird's repertoire.

Table 7: Vocalizations of Purple and Purple Rumped Sunbirds

Purple Rumped (January -May)	Purple January to May
March -April (first season female) freen freen	**May-August(first season male)** cheep cheep
Nov-Dec (male) i) cheleep cheelep ii) che che ple che ple pe pe pe che che ple che ple....	**January up toMay (f)** i) tchewe tchewe, twetchewe tchewere,twetchewere tchewere tchu tchu tchu ii) tchutchutchu iii) tchewet? tchewet? tchewet`
January-Feb(f) twe twe tweetweet, twee chewerrereruchewerrerreruchewerrreru twecheweerrt, chewera, chewera, twetweeru, twetweeru tweeweru, tweeweru twechewerr	tchewet`, tchewet? tchewet? tchewet` tchewet`, tchewet`, tchewet` tchewet, tchewet tchewet` tchewer` tchewer` tchewewr` tchewet tchewet tchewet! tchewet!tchewet! ...... twetchu twetchu .... tchewe, tchewe ...... tchutchutchu tchewe, tchutchu
July (season end) pepe...pee (m) tchu...tchutchu (f)	
	May (male) season end) i) cheeple, cheeple ii) ple ple che ple che ... iii) cheep

Grey Breasted Prinia

Prinias or wren warblers are a small sized 11-13 cm passerines widely distributed in the Indo-Malayan range. They are basically identified with dull rusty plumage and long black graduated white tail ending in a white tip. Prinias are always inducted into a bird community because they are skilled nest builders and never reuse their nests. Generally hanging ball shaped nests are woven out of fine grass fibres with several opening with a hatch. Other larger birds always have use for these discarded nests. The grey breasted prinia (*Prinia hodgsoni*) or Franklin's wren warbler is commonest of the prinias but they do not build such nests. They are leaf nest dwellers using saliva to gum the leaves similar to sunbirds and tailor bird. But the nests are built in the forks of shrubs that grow profusely and form thorny undergrowth of forests. Typically the prinias hold their tail straight up neatly with the feathers folded during breeding season but the grey breasted is usually found with the tail feather fanned out like the fantail warbler. In April one is likely to find a good sized flock of about seven members together in the undergrowth and medium sized shrubs all with their tails fanned out like this as if performing in a group ritual round their nesting territory.

Picture 34: Grey breasted Prinia (singing male)

The grey breasted is a colony breeder with a pair of birds - one female another male - performing the vocal and leading role. The prominent bluish grey breast patch that is generally identified with the species may be found only on the singing members. These two members are also the flock's breeding members.

Picture 35: Grey breasted Prinia leading female at breeding time

Strangely the rest are apparently nueters. It is important to observe that these two leading members do not form a mating pair and do not belong to the same generation. In a breeding cycle the male singer is a younger bird among them usually born to the female leading singer. The leading female singer pairs off with a previous season's lead male for breeding. The singing members are hatched separately and singly each in two different seasons so that the species has two breeding seasons in the same year. All the members in a flock participate in the reproduction even though there is only one successfully breeding pair. About five chicks may be produced in total in each of these seasons, January to April and May to August. The first season produces the next generation male singer and in the second the leading female is hatched.

This intergenerational leading pair keeps up sex differentiated solo song routines with the female producing peak vocalizations during the hatching-fledging stage of a year's first breeding season (her own) and the male in the same year's second season hatching-fledging stage (not his because he is not yet into breeding). Song notes are also strongly sex differentiated but the tenor of delivery remains the same, a sharp urgent persistent production of notes from edges of overhanging foliage. Both produce maximum vocalization in only one season each. In addition to the vocalizations produced by the singing members, the flock as a whole produces an intriguing visual display at the start of a breeding event in April. It involves the entire flock including the female lead singer who is into her second breeding event by now circumambulating the nesting site accompanied by the display of tails fanned out in a rhythmic and synchronized manner. This group ritual is carried out with a low buzzing (zeee..) and rustling (twirr...) sound. The leading female does not produce any vocalizations during this time, but one is bound to hear an older molting member (perhaps the previous year's leading female) while foraging in the undergrowth in a warbler-like fashion. She produces a two noted "vichuvichu" vocalization repetededly. The flock becomes invisible for a while and is out once again by late July and now is accompanied by newly fledged chicks. At this time the leading male produces his peak fully vocalized song – this time solo-display perched conspicuously on overhanging tree foliage. This singing is accompanied by jiggling his bulla conspicuously somewhat like jungle babbler males, displaying a puffed out throat from the top of overhanging branches. During such a delivery a whole retinue of newly fledged chicks and their parents can flit past quietly while the male makes a distracting show.

Table 8: Vocalization patterns of the Grey Breasted Prinia leading pair

Leading singing male	Leading singing female
Season initial (second of the same year (April)	Season initial (Nov -Dec)
	No vocalization
ze ´eee....	Hatching stage (first season of the same
e ´zeeze,e ´zeeze *etc.*	year (Feb to April)
e ´zeeeee......	twitwit, twitwit, twi
xze ´eee....	twitwitwitwitch, twitwi, twi?
eexzee, eexzee, eexzee, eexzee *etc.*	twitwi, twitwi, twitwi, twitwitwi, twi?
eexzeeeexzee, eexzeeeexzee, *etc.*	twitwitwi, twi?
twi$_{sss}$rra, twi$_{sss}$ra *etc.*	twitwitwitch, twitwi?
e ´eee.....	twitwitwitch, twittwit.......
e ´ve ´ee....	
e ´	Second season of the same year (April)*
e ´ ´eee...	vichu
e ´, *etc.*	vichuvichu,vichuvichu, *etc.*
	vichu, vichu, vichu)
Post fledging (second season of the same	vichve
year (July-August)	vichuvichu, vichuvichu *etc.*
dzerdzerdzer, *etc.*	
	(May)*
	vichuvichu, vichuvichu

Note: Grey breastd prinia has two breeding seasons in a year. November to April and May to August. The male and female of the leading / singing pair dominate in turn and produce vocalizations so that only one bird produces vocalizations in each of these seasons. *mark refers to an older bird likely last years leading female

It is possible that the ex-leading female is responsible for building the saliva gummed nest at the start of the season. She is found to be heavily molting and perhaps neutered due to the nest building chore. At least two nests - shallow saucer shaped - are built of different sizes and at different levels of shrub such as bamboo using different plant fibres. The nests may be located in diferent locations within the same territory. One of them is incubator for the next generation leading male and female. The other slightly larger and at a lower rung is a kind of communal incuabtor.

The song routine of the grey breasted prinia foregrounds several new aspects of bird song. Foremost among them is the question of who is the addressee. Not only do the lead singers among them achieve peak song in the post mating time and during in fledging, the song appears to be addressed pointedly to the just fledging chicks. In addition the singing male and female birds do not make up a mating pair but are from subsequent generations and have parent offspring relation. It is also significant that both male and female produce equal song (in terms of quantity and song types each has a minimum of two song types) even though the notes are sex differentiated. In addition to these issues it also raises the intriguing question as to the connection between the group synchronized fantail display of the flock and the vocalizations of the leading birds. The fantail display is highly ritualized and seems to be in response to the vocalization of the leading members – especially the younger just fledged leading male and seems to be an aspect of group cohesion.

3

Singing Nonpasserines

Cuckoos

The song of the cuckoo bird seems to be more the stuff of myth and poetry than real. It seems that there are about 20 races of cuckoo (cuculidea family) spread all over the world except in the Arctics but the fact is not all of them can sing and none of them are song learners. India boasts of a number of species only some are dear for their song, some others merely for whimsical ways and still others for fearsome beauty. The cuckoos are arboreal and inhabit the middle levels of trees along with the other passerines. But paleontology reveals that cuckoo birds must have evolved at an earlier date. Yet it is possible that the singers among them evolved alongside the passerines and that the song mechanism developed similarly to the song bird.

Among the cuckoo vocalists of an Indian forest we find the Asian koel's (Eudynamys scolopacea) sweet "kohoo kohoo". But he produces his song only for a brief period in April or May usually months of light showers and good weather perhaps explaining the association of the koel in the popular imagination with spring. The other singing cuckoo in the habitat is the bay banded cuckoo (Cacomantis sonnerati). It is a non resident bird in the region and arrives as if to confirm the monsoons well after the rains have set in. It has a distinctive song "wee laa la" unique and as musical as the koel's. One is more likely to hear the coucal's (Centropus sinensis sinensis and Centropus toulou bengalensis) variously pitched mysterious cuk cuk most of the months –there are two resident species in the same territory as the other cuckoos of their fami;ly - even though no match in song.

All cuckoo singers are seasonal performers, produce distinctive songs and do not show the capacity to learn songs. With the exception of the coucal female, which produces sex distinguishing call (a "krraahh" sound), only the males produce any kind of vocalization at all. It appears that the Asian koel sings only in his first season. This explains why the cuckoo song is so rarely heard. A young male still in the unpaired state produces the melodious song that the species is legendary for. This song possibly coincides with the birth of his female. Once he has paired off he does not produce any song and is a remarkably sedentary bird.

Field literature on cuckoos in the south Indian habitat list several different species such as Eurasian cuckoo, Indian cuckoo and Asian koel. Structural mimicry seems to dominate the intra-inter species dynamics among the cuckoos rather than song. And it is near impossible to make out the species composition in a population. They are strongly sexually polarized, and display sexualized morphology. They also strictly practice sex segregation, males of the population may be all found together on the same tree and their females on another. An Asian koel pair will never roost or perch together on the same tree. The male is strongly male and will not participate in nesting. Older Indian cuckoo (Cuculus micropterus micropterus) females are usually found living in segregation in some sort of seraglio on the same shrub as younger females of other cuckoo species such as Eurasian cuckoo (Cuculus canorus subtelephonus). All the females in the population display the typicl cuckoo female plumage morphology, but the shades ranging subtly from brown to deep brown to darkest black. Such intermediary plumage patterns are acquired and maintained by the practice of brood parasiting on diverse species, such as changeable hawk eagle or forest owl eagle.

Pictures 1a, b & c: Cuckoo intermediary morphological forms (female)

The cuckoos indulge in many paired ritual-like behaviour. Male Indian cuckoos are found to force- feed its younger males on berries that are addicting. Eurasian cuckoo pairs indulge in pair feeding of fruits like the papaya and can migrate many miles just to do that. Bathing is another ritual performed frequently I believe to ward off the after effect of molting. It is not uncommon to see the male sunning its wet feathers in the early morning after a bath in the river. Molting among most birds is an emotional affair when they lose their plumage patterns. The older

Indian cuckoo female becomes grousy and sulky. She also becomes tyrannical towards younger females. The Asian koel female will segregate in a grove at the margins of her eco niche.

The brood parasitism of many cuckoo species is a much researched. The cuckoo lays her egg in the nest of other birds far smaller in size like the warblers, flowerpecker and makes them raise the chick.Stranger still they can even use larger birds of prey such as the changeable hawk eagle (*Spizaetus cirrhatus*) as host to raise the young. Why do, let alone the little birds, even the large birds of prey tolerate the brood parasitism of cuckoos? How does this fit the theory of survival of

Picture 2: Koel (male)

the fittest? The changeable hawk is a typical preying bird. Even though he feeds on rodents and not birds, he is found to terrorize flocks of even large sized birds like cattle egrets. Entire flocks are found to take wing whenever he flies over their foraging ground. He emits a high shrill siren- like sound in warning as he flies over them and pelts stones kept in his gullet meant for grinding food. His nest is built in the spacious fork of large trees and is his natal home inherited from his parents. Such a nest can be passed over to a younger male after padding it once more and used again and again. Young fledged males just into their season and that have inherited the nest from parents can become host to cuckoos.

In the Bhadra area I found that a Eurasian cuckoo used the changeable hawk eagle nest to lay her egg and allowed the young bird to brood. Even though the new fledgling was then cared for by smaller birds that frequented the same tree such as drongo and myna. The question is how does the cuckoo manage to use him to brood?

First of all it may not be a coincidence the plumage pattern and size of the Eurasian female closely resembles that of the changeable hawk's. The newly hatched member was a female and was chequred like the changeable hawk. By this structural mimicry the cuckoo ensures that the bird does not see the difference. Secondly a possible explanation could be that the cuckoo sucks off the eggs of the changeable hawk so that when the chicks hatch the fledglings have imperfect crests. The crests on the crown of the changeable hawk are kind of feelers to sense even far off objects, losing it means loss of sensory perception. A bird that has

Picture 3: Pale Billed Flower Pecker commonly hosted by the cuckoo to raise its male chicks

his crest sucked off can be as deaf as ever and can be used happily as host. It is possible that this also makes him more tolerant towards other smaller birds such as drongo, myna, barbet even woodpecker who become the cuckoos allies in her brood parasitism.

Pictures 4a & b: Changeable Hawk Eagle& Eurasain Cuckoo female

Another clue to nature's toleration of the cuckoos' brood parasitism could be that it serves as mechanism for diversification of species. Darwin thought diversification as an important principle of evolutionary adaptation. Since the cuckoos depend on good populations of those birds which it parasites it doesn't serve its purpose to destroy the eggs or the fledgling of its hosts. Instead it can adapt them to its needs by feat of cosmetic surgery such as sucking on hatching eggs partially and injecting the sucked material into another. Perhaps this explains the existence of hybrid species such as drongo cuckoos, hawk cuckoo, cuckoo shrike so on. I believe cuckoo shrikes' beak is especially adapted to do this. Cuckoos are nature's agent of change. They enable birds to acquire suitable adaptations expediently.

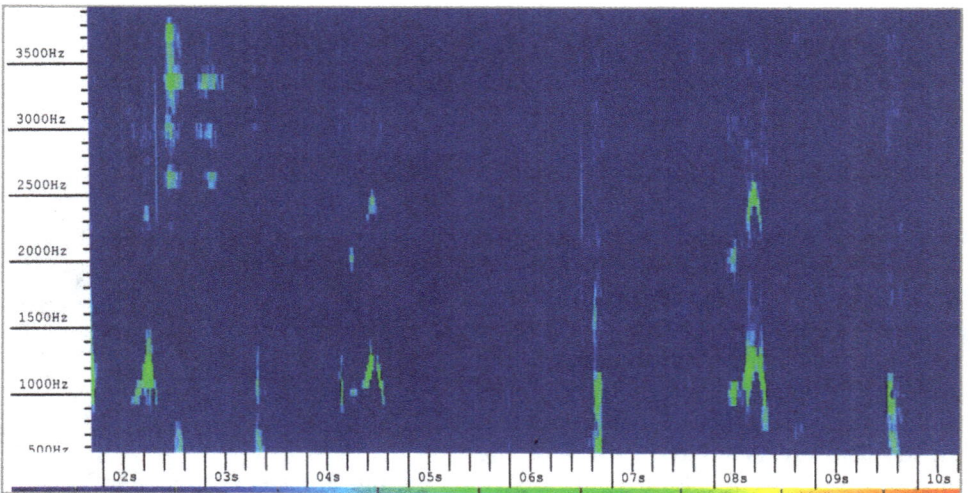

Figure 1: Spectrogram of the first song the cuckoo singer.

But all the explanations I have presented above merely refer to the proximate causes of the cuckoo brood parasitism. The question of course arises, is there an ultimate casue behind it? In other words, is it a genetically adapted characteristics and if so what lead to it? Indeed it appears that the cuckoos take recourse to brood parasitism in order to overcome a serious genetic handicap. In fact this handicap is similar to that experienced by the passerine song birds such as iora, magpie robin babbler and so on. I have presented an elaborate discussion of it in the concluding chapter of this book. Here it will suffice to say that the cuckoos parasite on other birds to raise their young in order to keep their song.

Blue Faced Malakoha

The Blue faced malakoha (Rhopodytes viridrostris) is a stunning singer that if one held a contest of song in a south Indian forest orchard perhaps the bird would emerge the winner. Among them both in the breeding pair can sing equally beautifully. They both would no doubt be rated highly over several aspects, for the variety of musical scales, command over vocal chord and voice, and expression and feeling. The female produces a song comparable to minstrelsy while the song of the male is comparable to a madrigal. And I make this comparison in all seriousness, such is their song capacity.

The malakoha belongs to the cuckoo family and is a local resident in the Bhadra region. Breeding season falls from July to November. Song can be heard

Picture 5: Blue Faced Malakoha (first season)

in November.They are found in a mixed flock with other song birds such as tree pie, racket tailed, orioles, flycatcher and babblers during this season. In keeping with the membership in the cuckoo family the malakoha reveal many of the typical

Picture 6: Blue Faced Malakoha (post breeding)

behaviours of the cuckoos. Like most cuckoos they prefer to roost on middle rung of thick foliage of wintering evergreen trees and seldom come to the canopies which are the feeding ground of sunbirds, mynas, and parrots in winter. They are as secretive and reticent as the Asian koel and parsimonious in the delivery of the magnificent songs. They resemble the coucals in the sex differentiation of their songs and also in the notes which is predominated by the "kra" and "kuk"syllables, albeit sung in a different dialect. But of course they have a far more skilled and elaborate song routine.

But the most intriguing of all is the brood parasitism displayed by them comparable to the Eurasian cuckoo. They host the nests of the Asian paradise flycatcher especially and engage the caring skills of the flycatcher female to fledge their young. The paradise flycatcher is a polygamist who has several females engaged in nest building and raising the clutch. But by the end of August –September the chicks are fledged and only require tutelage and grooming. At this time the flycatchers migrate to higher altitude cooler forests. Their tail streamers enable them to fly at high altitudes easily even if they are not equipped with strong wings. The female paradise flycatcher cannot keep pace with her migrating flock because her reproductive care has lost her flamboyant tail She is left behind in the nesting territory to fend for herself. But luckily for her she may have other females like her

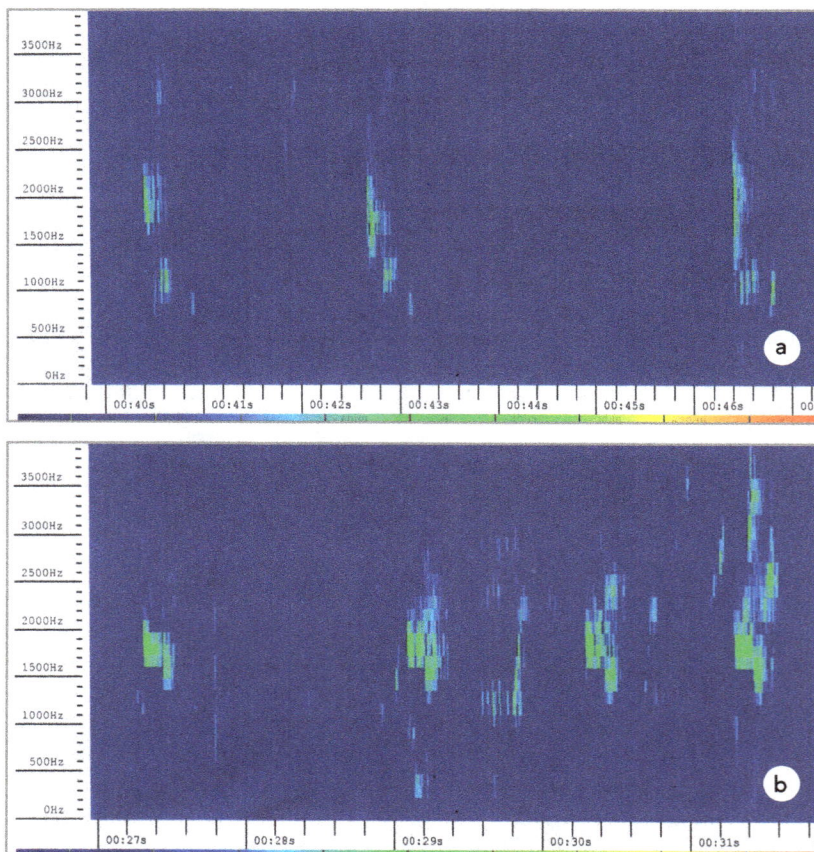

Figures 2a & b: Showing notes from the malakoha male (day four) & female (day six) duet song.

for company. Such is the reproductive demand on her. She is adept at building nest and hunting for grubs. In fact her nests are in demand among the hooded orioles too. The malakoha elicits her care and has his brood raised but under the keen eye of his female. Here is another instance of social altruism on the part of the paradise flycatcher female: she remains a willing captive through out.

The malakohas are not sexually distinguishable by morphology. The female is smaller in size and perhaps is a younger bird. But they are clearly marked by different songs. These different songs can be delivered at the same time from the same location or on separate days or time of day and different locations. When sung at the same time it can appear to be a duet. In the following it may be noted the day 1 to day 3 one note vocalizations of the male are repeated rhythmically for a considerable length of time and may be heard in the early morning and evening. The table suggests that malakohas are not a song learning species but perhaps one can hazard to say that the song notes show resemblances to coucals, crows and orioles.

Table 9: Sex role differentiated Songs of the Blue Faced Malakoha

	Male	Female
Day 1	` khk	**No vocalization**
Day 2	` khk	kraaun$_{kraaun}$kraaun$_{kraaun}$kraaun$_{kraaun}$kraaun$_{kraaun}$
Day 3	` khk	kraaun$_{kraaun}$kraaun$_{kraaun}$kraaun$_{kraaun}$ kraaun$_{kraaun}$
		kraaun?$_{kraaun?}$kraaun? $_{kraaun?}$kraaun? $_{kraaun?}$
	` khkiii	kraaun ?$_{kraaun?}$
		kraaun? $_{kraaun?}$kraaun?
		kraaun` $_{kraaun}$·kraaun`
	` krak	krukraaun`$_{kru\ kraaun}$·krukraaun` $_{krukraaun}$·
		kru kru kraaun`$_{kraaun}$·kraaun`
		kra kraaun`
Day 4	` khk	**same as day 3**
	` khkiii	
	` kruikk	
	` kraurrr	
Day 5	` khk ` khk	**same as day 3**
	` kruikk  ` kruikk	
	krako?krako? krako?	
	kruiko? Kruiko? Kruiko?	
	krakruyako? krakruyako?	
	` kraurrr kraurr	
	krikku	
	khko? khko?	
	krko? Krko?	
Day 6	` khk	kraaun kraaun kraaun
		kraaun` kraaun` kraaun` kraaun` kraaun` kraaun`
		kraaun` kraaun` kraaun` kraaun` kraaun` kraaun`
		kraaun` krayaro ` krayaro
		` kukrik ` kukirik` kukirik
		kraaun` kraaun` kraaun`kraaun` kraaun` kraaun`
		kraaun` kraaun`
Day 7	` khkiii	**no vocalizations produced**
	` khk	

White Cheeked and Coppersmith Barbets

The vocalization of the white cheeked barbet is a good example for the diversity of social uses a breeding song can be put to by the avian species. A breeding song is usually directed both at members of one's own flock as well as others in the habitat and forms the intra-inter species dynamics of a territory. But it appears that some species play a more prominent role in regard to the inter species life of an eco system. In the previous sections I have given an account of the mobbing displays of the red whiskered bulbuls in the presence of intruders, the tenor and intensity of the displays varying with the type of intrusion. Another example is the alarm call of the Tailor bird. Among this species the singing member is recruited to emit alarms for another member (such as young just fledged brahminy kite) in the territory during threatening situations perhaps because they are so willing to indulge in song type mismatching even as they show remarkable skill for mimicry. But the mobbing calls of the white cheeked barbet seem to be a self appointed one and very distinct in mode (tenor) of delivery from its breeding songs, even though the song notes are the same.

Picture 7: White Cheeked Barbet

The breeding seasons of the white cheeked barbet (*Megalaima viridis*) and coppersmith barbet (*Megalaima haemacephala*) fall from Feb to July and Jan to May respectively. This is the time maximum vocalizations are produced by both species. Adult members of the white cheeked can be found in identical looking pairs in March. The male produces a vocalization in the presence of his female which appears to be making a ceremonious announcement of her identity to the community. Full song capacity is displayed by May -July when the chicks are fledging under the care of the female and she is invisible in the territory.The song then develops different notes suitable to the development of the hatchlings and the proceedings of the season. The mobbing calls (see spectrogram given in Chapter one) are a variation of these songs and mark the regular activities of the day, movements of people, animal or bird. These mobbing calls can vary in intensity and tenor depending on the stimuli that provoked it; a routine intruder is received with a routine delivery, a new or a serious threat at a higher intensity of notes.

The coppersmith that shares the territory has a different song routine. Its songs are strictly seasonal. Unlike the white cheeked a young coppersmith in his first season can be heard intermittently as early as October; a one note vocalization is produced at this time rather than song. Peak singing can be had only for a short period past peak breeding but during fledging that is April-May. Capacity for song

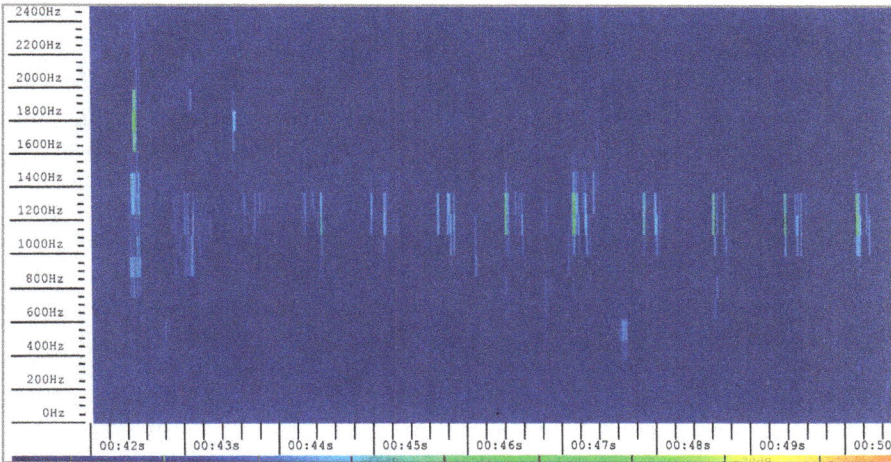

Figure 3: Female identity announcing song of the white cheeked barbet

is achieved in January the start of the nesting season. Rest of the year one can hear only the white cheeked.The songs of both the barbets are made of one note but sung out in various combinations and tonal variations to render a verse song out of it. Dramatic variations in pitch and rhythm are also used to produce this clever expressive composition by both species.

The different song distribution between the coppersmith and white cheeked is matched by the different types of reproductive participation they engage in. First of all it appears that in a season the coppersmith clutch is of a larger size with at least three eggs laid at once and fledged together. The white cheeked produces a clutch of two eggs but laid one by one with a time interval in the same nest hole, so that the newly hatched first chick aids the incubation of the second. The coppersmith male has bigger share in the breeding process; He is found diligently lining the brooding nest and feeding his sitting female during this time. He takes over all post incubation duties which makes him totally invisible and silent in the territory after the initial display singing. Not so the white cheeked male who has marginal share in the actual breeding process. He of course contributes to the making of the tree hole nest. But his contribution is no more than this. Tree holes are made suitably in a spot that receives plenty of sunlight. Once the eggs are laid, his female seals up this nest with

Picture 8: A coppersmith barbet steals away with a flower bud in March to line the nest of his sitting female

fluff produced out of her saliva. This method of incubation frees her to feed by herself during this time. She is also entirely in charge of the fledging. He is free to sing all the year through.

The songs of the two species of barbet that inhabit the same territory for the purpose of breeding as well as foraging does not exhibit any point of contact or conflict. There appears to be peaceable cohabitation of the two genetically related species. Other woodpeckers like the flame back, white napped and brown rumped found in large colonies of six to ten birds in the same territory do not indulge in any kind of song trafficking with their cousin barbets. Nor do they have any elaborate songs for that matter.As far as intergenerational continuity is concerned only one barbet is heard in a territory at a time, even when the population is more than one pair. It is possible that a barbet pair has only one breeding season if it is successfully delivered. Older pairs lose song capacity after their first season and are neutered. Since the songs are typical even though elaborate there is no question of song learning. The females of both species do not produce any vocalizations.

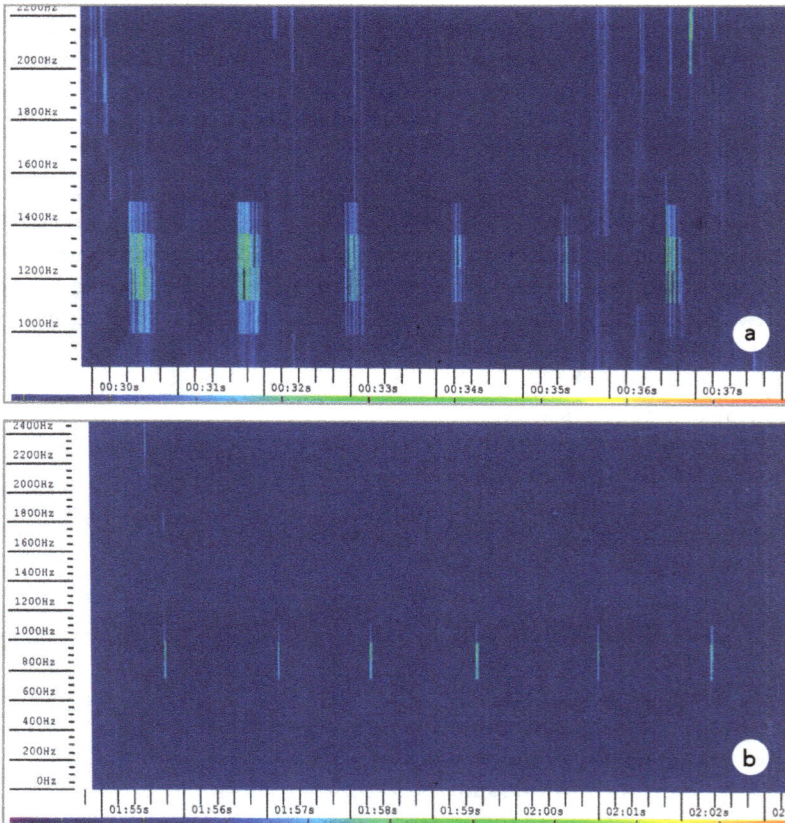

Figures 4a & b: Spectrograms of portions of the white cheeked and copper smith barbet's territory monitoring songs

Table 10: Vocalization patterns of the White Cheeked and Copper smith Barbets

White cheeked (Feb-July)	Coppersmith (Jan- May)
no vocalization (Imfl July- Jan)	cuk (Imfl Sep- Dec)
Season initial (bm)	Season initial(bm)
no vocalization	Cuk, cuk,cuk,...
Hatching (bm)	Hatching (bm)
kor, kor,...	Cuk,cuk,cuk....
kocor, kocor...	Cuk,cu,cuk,cu...
kocarkocar?...	Cukcu,cukcu...
kocorokocoro....	Cucuk,cucuk....
kor, kocor, kocorkocora....	Cu,cuk,cu,cuk...
kocoro,kocoro...	Cuk,cuk,cuk....
kocor, kocor, kocoar?(3times)	.....
kocar, kocar, kocar!	Post hatching(bm)
Post hatching(bm)	no vocalization
Kocorow$_{asss,}$ kocor, kocor, kocor	Post fledging(bm)
Kocorcorawa, kokcro, kokcro	no vocalization
Post fledging(bm)	
Kooocorawa co co co	
Koooooooooooorkococro,kococro,kococro,kococro...	
kococroooowkocrkocrkocr...	
season end(bm)	
kocrowakocrkocrkocro	

Legend: bm=breeding male'immfl=immature fledgling, feb to July and Jan to May are the respective breeding seasons.

Spotted Dove

If you fancy ideal life as that in which quiet mellow days of winter and spring are spent to the sweet spotted turtle dove crooning of a hundred tales whole day long just out side your bedroom window then you must come to the South Indian country side in winter. History will tell you that this is life fit for a Moghul. The turtle dove was a familiar in the zenana, its soft cocorcocor coor coor a sapient to

Picture 9: Spotted Dove on bamboo branch

fraught nerves of an overcrowded household. In many a murals and fracas a pair of turtle doves symbolize steadfast love and permanence. Doves are perhaps the most written about among birds and the most historical because of their use as carriers. References to them I believe go back even to the earliest civilization such as the Mohenjadaro Harappan.

Typically all doves are forest dwellers and shy of humans. Only the Spotted dove (*Streptophelia chinensis suratensis*) and rock pigeon are found near human habitations the others warily prefer to live in forest clearings and edges away from the reach of humans. The spotted dove lives in ever green trees but during the breeding season, late winter to early monsoon pairs forage on the ground. Obviously breeding demands that the food intake of the bird is greater and they alight to the ground to pick at grains and seeds found in grassy lands and fields. The breeding dove produces milk with which to feed the chicks. The singing bird once released of his duties towards his brooding female continues to sit in the middle rung of partially shed trees coor cooring sweet tales of gone by seasons rest of his life

In the popular imagination mating doves are associated with fidelity and constancy. There is much literature on the courtship behaviour of turtle doves and this is attributed to the emotional bonding between a mating pair. Contrary to this field observation suggests that the male dove is very fickle. His capacity for song makes him indifferent to the reproductive needs of his female and he is known to desert the incubating female.Among them the female takes over all the reproductive responsibilities and this leaves her vulnerable especially during the incubation period when she cannot afford to leave the nest of eggs. What we think as attachment such as courtship feeding appears to be the female's attempt to achieve secondary imprinting on her male so as to have someone to feed her during the crucial period. There is always a chance for the male to abandon her for a song. The female of course forfeits her song for the sake of perpetuation of the species. Only the males produce songs. It is my suggestion that sexual difference among many bird species (especially non passerines) is not mandatory, coming into play only during the breeding season. It is possible that in most cases sexual polarization is determined by the differences in the age of the breeding pair. Among doves sexual difference is correlated to the age of the mating birds, the younger is always female.

Most doves can be distinguished by their different calls. The exotic Emerald dove (Chalcophaps indica) has a two note cococoocoo signature call, the Laughing dove (Streptophelia senegalensis) makes a coughing cococo, the Rock pigeon (*Columbia livia*) simply coors and the Yellow footed green pigeon (*Treron pheonicopatra*) makes a low pitched sonourous moaning cooocooo hiding among the foliage.

Picture 10: Yellow Footed Dove

Table 11: Vocalization pattern of the male Spotted Dove

Ist season	Pre mating / Mating	Jun -Aug	No vocalization
	Hatching	Sept-Oct	coor coor
	Fledging	Nov-Dec	coco coor coor { cocorco coor coor cocorcoor coor....
	Post fledging	Jan-March	coco coco coor coor coco coco coor coor cocorcocorco coor coor cocorcoco coor coor...
IInd	Pre- mating	April	coor coor
	Hatching	Sept-Oct	coor coor coco co coor coor cocorco co coor coor cocorco coor coor cococococo
	Fledging	Nov Dec	coorco? coco coor coor ? coorco? coor coor? cocor co coor coor
	Post-fledging	January	corco coor co coco coor coco coco cococo co coco coco co ccococo coco co cococo coco co cococor coco cococo cor cocococo cococor cococococo corco co co? cocor coco? cococo?.....

The songs of the spotted turtle dove follow an interesting curious routine. Vocalization is not produced until the male bird in his first season enters into a definitive breeding relation, which takes place in May-June. The first song is produced when the eggs are laid and the female is incubating. This coorcoor song is kept up whole day in early winter paused only for the bird to feed his incubating female. By November this song has changed to a longer coco coor coor coor produced continously. Subsequently another new note (cor) is added cocorco coor coor. The song develops into a full length verse composition in this manner. By early January the bird is in full possession of his songs. They are composed of all the verse lines developed earlier with increased numbers of 'co'. All of the lines are produced in

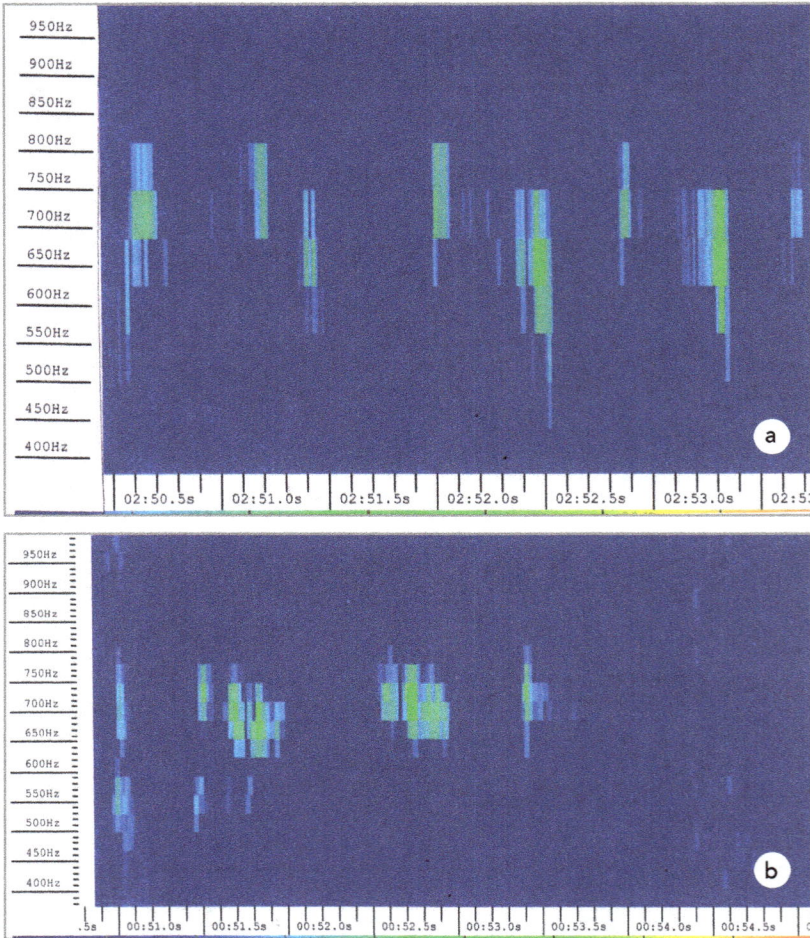

Figures 5a & b: A comparison of portions ("cocro co co co" & "cocro coor, cocro co coor") from a dove's post breeding songs produced in two consecutive seasons

various sequences as a one continuous song. Accompanying these developments there is also a change in delivery becoming softer and mellufluious. I suggest that the development of the song encodes the breeding status of the bird and imparts information of the sex of the newly hatched egg. The numbers of the 'co' sound also seem to indicate the breeding age or his season. Older birds thus produce longer and longer songs. When the season has been particularly good there can be several such successful males cooring away on different trees in the same territory. Each male's song distinguishes him according to age and status.Throughout these deliveries the bird can take breaks to visit his female and her chicks.There is no question of rivalry or of song learning among them.

4

Why Do Birds Sing?

"...And be the singing masters of my soul Consume my heart away..."

(Sailing to Byzantium, W B Yeats)

In this concluding chapter I shall review what I have been saying so far and attempt to uncover a rationale based on some accepted general principles to justify for having interpreted my empirical observations the way I do. Any observed behavoiur should have an explanation in ultimate causes (Alcock, 2005). Therein lies the proof of an explanatory framework. The subject of this study has been songs of bird. There are actually two questions that I started with; one the big question why do birds sing? The second one of equal significance is why it is that capacity for song is unequally distributed among birds, why is that only some birds can sing even when of the same genetic make up? A third question has begun to pose, why is there diversity in song type? Or what does it mean when two or more birds sing differently between them?

In ornithological literature generalization on the bird song have always stressed their relation to reproduction. Songs have been attributed to breeding male bird and his sexual prowess. Let us examine this generalization: It is clear that it is based on a priori assumption of a nuclear paired reproductive unit among the birds, derived from the fact that birds are biologically bi parenting. There are two principles that underly this assumption on songs of birds: One is that biparenting determines the species identity/ genetic make up of an organism and is its sufficient condition; and a second derived from the first, the assumption about the social organization of reproduction as a paired unit. Even though it will appear that biparenting determines the genetic specificity of birds because the new member is produced by the mating of only two individuals it is not a sufficient condition to understand genetic specificity as we shall examine later in this chapter. For now I shall focus on the organization of reproduction as a paired unit. My empirical field study of birds in their natural habitat points out the error of this assumption. It seems to me that this has been extrapolated from our own human - perhaps peculiarly modern day—normative reproductive arrangement of the nuclear family.In their natural setting most bird species do not engage in paired reproduction instead breeding

Picture 1: Oriental white eye flock

is a coalition among several birds. Many birds like the iora, robin, wren warbler, and tailor bird form a reproductive coalition of three or more birds. Such a breeding junta invariably has a role for a singing male but this singer most of the time is not the breeding mate. In fact he is usually a bird who has achieved sexual maturity but since he has biologically opted to be singer falls behind in mating. My hunch is that a bird that is invested in singing, especially those which produce sustained singing all its life is forced into a biological compromise and relinquishes his privileges as mate. So in a breeding junta the breeding male is very rarely the singing male. There is a lead singer and a lead breeder.

My study of birds of the Bhadra region presents us with broadly seven different types of sexual reproductive arrangements or reproductive systems (Halliday, 1980): represented by oriole, magpie robin, flycatcher, iora, bulbul, wren warbler and dove. It may be noted I have not included cuckoo in this list and may be considered separately. These seven different systems might appear diverse and uncomparable, but the truth is all are governed by the same principle, where the reproductive roles are distributed among more than the mating pairs. The only exception is the pair breeding group orioles, sunbirds and barbets, *etc.* We find that among birds that do not produce sustained singing such as hooded oriole and coppersmith barbet the breeding bird is the singing bird. But he is a seasonal singer who does not produce his songs after a brief period season initially. Peak song is produced after the mating and before the eggs hatched. Rest of the time he is either silent or can only croak. Such birds also practice purely pair breeding and the male has a larger role to play in the reproductive process. He contributes to the fledging of the chicks and therefore cannot be heard once the eggs are hatched. In both instance females produce minimal songs and that before mating. Among pair breeders such as white cheeked barbet and sunbirds, that produce a great deal of vocalization during season they do not sit on the eggs but have sealed incubating nest type.

Magpie robin and iora have similar type of reproductive arrangement. In both the species the singing bird is not the breeding and the reproductive arrangement involves more than the actual breeding pair. Among ioras we find that the singing male bird totally forfeits his

Picture 2: Newly fledged Babbler Clutch

reproductive role and becomes part of a reproductive flock with more than two or more breeding pairs. But he remains a sustained singer that continues to sing even after the season is over (round the year) albeit a poorer version of his season's best. We also find that his female who contributes to the fledging of the chicks and the breeding pair produce minimal song at the beginning of the season only.

The singing magpie robin is inducted into a polyandrous trio in which he is not the breeding male but plays a surrogate role. He produces peak song starting at mating time up till the chicks are hatched or in some seasons till fledged. The breeding male magpie does not produce song comparable to the non breeding male. His songs are limited to his first season before he enters definitively into the breeding coalition as breeding male.

The Tickell's blue flycatcher on the other hand can sustain his song all the year through without seasonal changes. He is the only singer who fits the classical definition of song bird: He starts to sing in his first season even before pairing and continues to sing even after his male offspring(s) have learnt their song. Collaterally to this we find his reproductive participation is limited to the period of mate selection when he is found contributing extensively to the fledging of his female and not his own off springs. This is the only time he remains silent in all his adult life. Since the male seeking a mate collaborates with the actual breeding pair, contributing immensely to the sitting and fledging. The actual breeding pair is freed of these duties and has all the time to take over to song. They show equal capacity as well as equal use of song. Comparably among magpie robins both the breeding male and female sing only in their unpaired state from the period between fledging and uptill reaching breeding maturity. Among the Tickell's the female loses the species typical song once the eggs are laid (like the female magpie) but she takes up night time singing. This night time singing performs the same function as the male's day time singing: safeguarding the territory from unwanted intrusion. Her song capacity is comparable to the magpie robin breeding male's and female's: she sings in her first season before definitively taking on her breeding role.

The bulbuls and tree pies display another version of the reproductive coalition in which an older pair takes over the privilege of fledging the chicks of a younger breeding pair. This correlates with the sustained song capacity of the breeding pair. Among other passerines observed in this study tailor birds engage in a kind of polygamy with two females attached to a male. One of the females is a neutered reproductively inviable but contributes to the raising of the hatchlings. Differences in phenotypes and reproductive capacity are produced by changed allopatric conditions while fledging at the stage of chick. Tailor bird therefore can be grouped with magpie robins.

Among the doves our sixth type the breeding male produces song only after the mating and reaches his peak capacity after the chicks are hatched. After this he keeps up steady song. This may be accounted by the role of the singing male in feeding the brooding female until the chicks hatch. Once they are out of the egg their fledging is her sole responsibility.

Pictures 3a & b: Wild Grey Jungle Fowl & Red (domesticated) Jungle Fowl in fully developed post breeding plumage. The former native to Bhadra forest is as flamboyant as the peacock

Lastly we have the group representing the truly cooperative colony nesting passerines such as thrush, wren warblers and babblers. These birds are always found in a sizeable flock but usually there is only one breeding pair. These are also the lead singers so that there is male and a female singer. The rest of flock is made up of neutered members who play a helping role. Among babblers the vocalizing male is a younger member in his first season who plays surrogate to his breeding flock. But like the magpie breeding male he does not produce any vocalization once he enters breeding in subsequent seasons. Similarly the grey breasted prinia, a male in his first season plays the lead singer role and then gives up when entering breeding. But among these the female also produces song and somewhat like the Tickell's female during her first breeding season. So that she is both the breeding female and the lead singer. Among thrush we have a migrating species but this does not deter them from singing during breeding season. They are comparable to the grey breasted prinia. Among the orange headed thrush for *e.g.,* there is a pair of singing birds while the other members do not produce song.

Table 12a: Song Distribution and Quantity of Song Produced over Two Breeding seasons For Passerines

Reproductive stages in connection with breeding fe/males and singing male (if applicable)	Scimitar babbler* (May to Dec)		Asian brown (Nov to May)		Tickells (June to Nov)		Magpie Robin (March to Sept)		Reproductive stages in connection with breeding fe/male and singing male (if applicable)	Iora* (Feb to July)			Black hooded Oriole* (April to Sept)		Tree Piie (Pallida) (April to July)		Tailor bird (June to Nov)	
	M	F	M	F	M	F	BM S/M	F		B/M	SM	F	M	F	M	F	BM/B FM	S/FM
1st season																		
Pre-pairing	N	Y	N	N	Y	Y	Y	N	Pre-pairing	Y	Y	Y	Y	N	Y	Y	N	Y
						Y	N	Y	Pre-mating†	Y	Y	Y	N	N	Y	Y	N	Y
						Y	Y	Y	Post-mating†	N	Y	Y	N	N	Y	Y	N	Y
						Y	Y	N	Hatching	N	N	N	N	N	Y	Y	N	Y
						N	N	N	Pre-fledging	N	N	N	N	Y	N	Y	N	Y
						N	N	N	Post-fledging	Y	Y	Y	Y	N	N	Y	N	Y

Contd...

Table 12a Contd...

IInd season											NDA			
Pre-mating	Y	N	N	Y	N	N	N	Y	Y	Y	NDA	N	N	N	Y
Post-mating	Y	N	Y	Y	Y	N	N	N	Y	N		N	N	N	N
Hatching	Y	N	Y	Y	Y	N	N	N	Y	N		N	N	N	N
Pre-fledging	N	N	Y	N	Y	Y	N	N	Y	N		N	N	N	N
Post-fledging	N	N	N	Y	Y	N	N	N	Y	N		Y	Y	Y	N

Legend:*Female exhibits displays season initially or end.

† in the case of non breeding male/non breeding singing male pre& post mating will refer to the mating of the breeding pair.

M/BM =breeding male; F=breeding female; SM= non breeding singing male

SFM=non breeding singing female; NB=nueterd members; NDA= no data available.

Note: Ist season and IInd season refers to the breeding age of the birds. In case of magpie robins the Ist season of the SM coincides with the II season of the BM. In case of Tickell's female her Ist season coincides with IInd season of her male. In case of scimitar babbler an unpaired female's Ist season coincides with a BM's IInd season so that vocalizations are produced in the same time period. So the table must be read keeping this in mind.

Table 12b: Song Distribution and Quantity of Song Produced overTwo Breeding seasons For Passerines

	Bulbuls						Leaf birds										Orange Headed thrush (April to Sept.)		Sunbird				Grey breasted Prinia (March to August)			Reproductive stages in connection with breeding (fe)male and singing (fe)male if applicable
	Red Whiskered		White browed*		Red vented		Blue winged* (April to Aug.)			A. Frontalis (Mar. to July)			NSR (July to Dec.)	A. Insulais (Dec to Mar.)					Purple rumpd Nov-Mar		Purple (Jan-May)					
Iˢᵗ Season	M	F	M	F	M	F	M	F	NB	M	F	NB	NB	M	F	M	F	M	F	M	F	M	F	NB	
	N	N	N	N	N	N	Y	Y	Y	Y	Y	Y	Y	Y	N	Y	N	Y	N	N	Y	N	N	N	Pre-mating
	Y	Y	Y	N	Y	N	Y	Y	Y	Y	Y	Y	Y	Y	N	Y	N	Y	N	N	Y	N	N	N	Post-mating
	Y	Y	Y	N	Y	N	Y	Y	Y	Y	Y	Y	Y	Y	N	Y	N	Y	N	N	Y	N	N	N	Hatching
	Y	Y	Y	N	Y	N	Y	Y	Y	Y	Y	Y	Y	Y	N	N	Y	N	Y	N	Y	N	N	N	Pre-fledging
	Y	Y	Y	N	N	N	N	N	N	N	N	N	N	Y	N	N	Y	N	Y	Y	Y	Y	N	N	Post-fledging
IIⁿᵈ Season	NDA		NDA				NDA									NDA		NDA				NDA			

Table 13: Song Distribution and Quantity of Song Produced in One Breeding Season For Non Passerines

Reproductive Stages		Barbet			Spotted Dove May-Nov		Cuckoo					
		White cheeked Feb-July		Copper smith Jan-May				Malakoha (Nov to Jan)		Greater Coucal May- Nov		Asian Koel April-June
		M	F	M	F	M	F	M	F	M	F	M	F
Iˢᵗ S E A S O N	Pre-mating	N	N	Y	N	N	N	N	N	Y	N		
	Post-mating	Y	N	Y	N	N	N	N	N	Y	N		
	Hatching	Y	N	N	N	Y	N	Y	Y	N	N	Y	N
	Pre-fledging	Y	N	N	N	Y	N	Y	Y	N	N		
	Post- fledging	Y	N	N	N	Y	N	Y	Y	N	Y		

Note: Koel male produces songs only in the pre pairing stage in his first season. Among malakoha the juvenile male and female produces a one note vocalization briefly in January- February.

Empirically there is no doubt that capacity for song is inversely related to the extent of participation of a bird in the whole reproductive process which involves nest building, mating, incubating, fledging and tutelage of the juvenile adult(s). Lesser the participation in this process greater is the song capacity of the male. This can also account for sex distribution of songs but only partly: since among most birds the female's participation in the reproductive process is greater than the males such as doves, white cheeked barbet, and flycatcher so on. Among many other this contribution is equal, such as coppersmith barbet, and iora. In the former cases the role of the male bird maybe limited to feeding his brooding female when she cannot get out of her nest. Among all these birds we find that females either do not sing or produce a feeble song only season initially.

Even though our field observations have revealed that there is a predominance of male lead singers among the different bird species. There is good evidence that the above rule is even more binding on the female should she undertake the singing role. In the Bhadra region we find two distinctive cases of female lead singer: among sunbirds and tailor birds. The female singing tailor bird fails to produce viable eggs and has to give up her breeding role to a younger female. Among both purple and purple rumped sunbirds the female produces an elaborate song all through the breeding season but then she uses leaf incubators to which her male contributes a great deal. Other passerines with females that share the role of lead singer such as Tickell's blue flycatcher, grey breasted prinia, orange headed thrush, even babblers are comparable in that they all have surrogates inducted into the breeding coalition. Among the non passerine cuckoo- like brood parasiting malakoha both in the breeding pair show equally elaborate song capacity, even though the song notes and patterns are different between them.

Yet I suggest that reproductive participation cannot completely explain the sex distribution of song. For instance the surrogate role of the magpie robin is comparable to that of female of the singing male iora. Then why does the iora female not produce song like him or like her male? Likewise a question may be raised why is that the hooded oriole female does not replace her male and take to singing even as he replaces her on the reproductive front and quiets up his singing. The Tickell's flycatcher female of every alternate season is both a phenomenal singer as well as produces a clutch of three eggs. But why is it males in her clan do not lose their species song like she does and produce a learnt song (imitative of sounds in their environment) after season initial like she does? There must be a deeper mechanism than individual reproductive functionalism underlying sex song distribution. We shall try to understand this later.

In the same light let us review distribution of song among same species male members especially intergenerational continuity of song. By intergenerational continuity of song I mean the passing down of the species specific song. Most birds of course have a component that is made up of imitation of other birds in the habitat. In some this component may be greater. But in all the bird songs there is a fixed species specific component. This component may be relating to sound units, song pattern and song routine. In current ornithology (see Alcock, 2005) it is thought that this species specific song may be handed down genetically. Perhaps one should add the capacity to learn new songs also. Even though it is baffling why not all the members of a species can sing. Intergenerational continuity seems to be important to our understanding especially with regard to the species typical song: for a study such as this could throw light on the nature of genetic specificity of birds and their song, genetic divergence and the study of speciation. Field observations on intergenerational continuity, as I have argued in my sections on iora and magpie robin, suggest that birds do not learn to sing their species specific song. All the members of a species inherit their song. But what is the nature of this inheritance and how it is that only the lead singer vocalizes - this I shall explore later in my discussion on the development of song mechanism in avis class and the notion of song suppression further in this chapter. The data and family trees for intergenerational song continuity I present in the following analysis are primarily based on visual head counts of the birds and my familiarity with them in the territory. I have also relied upon data such as average size of clutch and breeding season *etc.*, taken from Salim Ali ten volume study (Ripley & Ali, 2001). In almost all the instances my observations are without exception corroborated by Ali's records. So I am encouraged to present them here. It is also important to remind here that the trees do not give any information on loss of eggs or hatchling, which is always the case.

Only among the Tickell's do we find any clear intergenerational continuity of song. Among them song capacity is carried down from father to multiple offspring and in addition both father and offsprings keep their song routines in the same territory at the same time. They do not display any rivalry or contest but co-exist smoothly sharing the territory for song performance and breeding both. That is why they grow into a clan. Among them the breeding male/female is also the singing bird. Combined with this equal song capacity among the older as well younger male, and female Tickell's, we find equal distribution of reproductive responsibility, an arrangement which does not show any sexual polarization in contribution to reproduction including territorial duties. The female's night time singing has the same function as the male's day time song routines; ensuring that the territory stays monitored. This can be arduous because up to three eggs can be had in a season. Eggs are produced with a time interval and it is possible laid in different nests. So the entire process is lengthened.

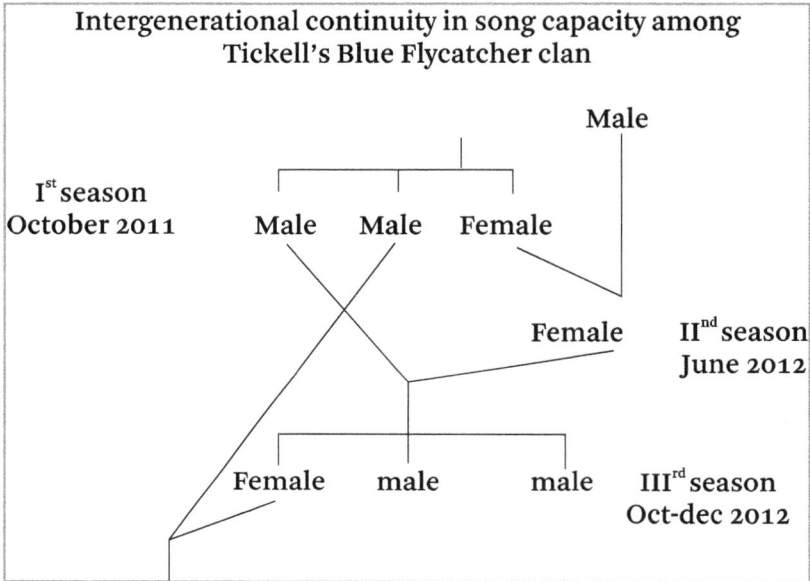

Intergenerational continuity in song capacity among Tickell's Blue Flycatcher clan

Male

Ist season
October 2011   Male   Male   Female

Female   IInd season
June 2012

Female   male   male   IIIrd season
Oct-dec 2012

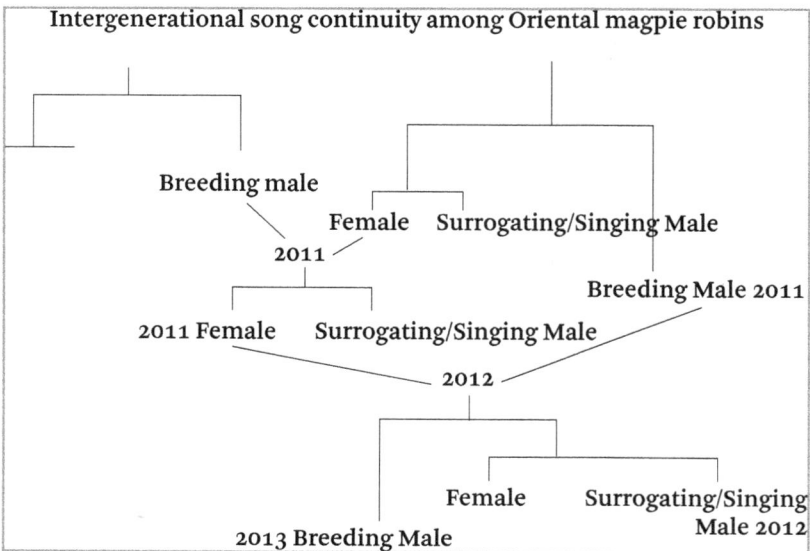

Intergenerational song continuity among Oriental magpie robins

Breeding male

Female   Surrogating/Singing Male
2011

Breeding Male 2011

2011 Female   Surrogating/Singing Male

2012

Female   Surrogating/Singing
Male 2012

2013 Breeding Male

Figures 1 & 2: Tickell's family tree & Magpie Robin

We would expect a comparison between the Tickell's and magpie robin since they are grouped under the same family. But we find that they do not have same kinds of reproductive arrangement or intergenerational song continuity. Magpie robins do not pass on equal song capacity regularly down from one generation to another as is clear by the presence of a breeding male and a surrogating singer male. But they do show an interesting variation of song continuity that elicits comparison with Tickells'. All the males are singers in their first season but each bird produces different quantity of songs according to the role they play in the polyandrous reproduction junta, breeding or surrogate. The breeding male can sing only in his unpaired state before he enters definitively the breeding relation. Only the surrogate can sustain his song in his subsequent seasons. The surrogate's song is more elaborate than the breeding male's, he is always younger and is from the same clutch as the breeding female. Thus song capacity (and reproductive role) varies between two consequent generation and shows similar continuity between alternate generations.

Next, among the ioras the intergenerational continuity in song capacity is a broken one. A bird from every alternate season turns out to be the singer; likewise the breeding male. The lead singer is not the breeding male and they do not belong to the same generation. The family tree for ioras is as shown below. The breeding span of an iora is for two seasons and a pair of eggs is laid every time. Songs are passed down not so much from father to offspring but from uncle to nephew (to put it in human parlance) every alternative generation. Therefore at a given time there is only one singer in a territory.

Intergenerational Song Continuity among Iora

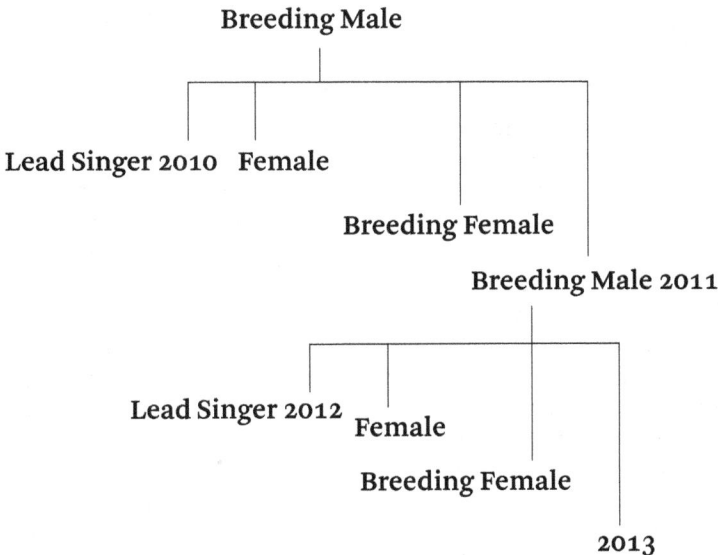

Figure 3: Iora family tree

The other passerines studied in the territory, racket tailed drongo, leafbirds, babblers, prinia, sunbirds, treepies and orioles all of them do not show any intergenerational variation in song capacity. But among the former three there is always only one lead singer (pair) in every generation clutch. Among birds that produce sex differentiated song type and capacity the intergenerational continuity is alternating for both song types. Sunbirds and prinias display such sex distribution of song; not so the racket tailed drongo and leafbirds. Among the bulbuls the red whiskered show song type variation among members in the territory even though not in song capacity as such. This variation can be allopatric as well as intergenerational and arises mainly due to differences in bird population in the respective territories. A generation of red whiskered bulbuls that learn the song of their own subspecies red vented fail to produce their own species typical song. On the other hand the other two species do not show any variation or learning. The intergenerational variation of song among leaf birds is allopatric but the species typical song remains invariant element of their song. In addition sympatric members produce reverse song type matching. Similarly the song repertoire matching of the tailor bird, we know that such a singer bird is inducted into a breeding coalition as well as into the territory's community of birds. Field observations point strongly that there is continuity in this role between generations. A bird in such a role keeps it for only two seasons and is replaced after that by a younger bird who functions exactly like him.

Among the non passerines spotted doves show as consistent song capacity across generations as do Tickell's among passerines. Several adult male doves can be found singing simultaneously in the territory from different trees. The older bird in such a covey can be identified by his more and more elaborate song with the passing of each season. White cheeked barbets we find that there is only one singing male even if there are many pairs in the territory. But I do not have sufficient data to analyze this phenomenon. I suggest more field work ought to be conducted before interpreting them. The malakoha like the cuckoo have intergenerationally invariant song pattern. Songs are strongly sex differentiated but not in song capacity.

The above discussion has been on the intergenerational continuity of song. It does not give an account of the actual process of transmission of song between the generations. To begin with I have already noted that the species specific component may be inherited rather than learnt by tutelage. The learning stage in all the birds under observation for this study occurred only upon reaching complete adulthood (usually signified by success in breeding). And usually is accompanied by imitative learning of other bird songs and sounds in the immediate habitat. The complete version of the species typical song is produced long before this learning stage early in the growing juvenile lead singer among most species, such as iora, Tickell's, tailor bird, leaf bird, king drongo and magpie robin. The magpie robin breeding male produces the species typical till he reaches breeding capacity but loses it by the time he enters into a definitive mating relation. It is also significant that among birds such as tree pie, oriole, sunbird, barbet and dove that do not show any evidence of song learning but invariantly produce species typical song, the full version of the species song is not achieved until reproductive success is clear. This could possible explain why they do not learn any new song because the species song is delayed.

In those birds which show capacity for song learning the species typical song was completely replaced by imitative songs in the completely adult bird, e.g Tailor bird. Among red whiskered bulbuls this phenomenon of erasing the memory of species typical and replacing it with learnt song happens to subsequent generations rather than in the same adult. They are found to learn species typical song of other species of bulbuls easily; when this happens that generation never produces the species typical song. Among Tickell's flycatcher younger birds always produce purely species typical song in about two months from fledging the female even sooner than that. Learnt songs are produced only upon confirmation of the breeding relation. Thus the female switches to night time repertoire singing after she starts to produce her first season eggs. In the Tickell male the species typical song is not replaced but merely enhanced by addition of learnt units; this same occurs among leaf bird and king drongo.

Having reviewed my field observations let us return to my examination of the logic of the generalization that invests both singing and breeding in the same bird. In the natural setting we know that reproduction among birds is an all consuming and serious venture that requires a little bird's full energies. It involves collecting of nesting materials, building of nest,

Picture 4: Pied bush chat and fledgling

mating, incubating, and fledging of the chicks. Birds that produce a clutch of eggs can have them each egg with a time interval so that the entire process is prolonged. In addition, once fledged the immature adults undergo a period of tutelage as ward of an adult. Among most birds this time period after fledging and before breeding is of crucial importance to species survival when the young ward learns to locate food resource, nesting sites and materials, as well as learn to recognize the species song.

In view of this it seems illogical and uneconomical to think that a bird can both song all his life time and still contribute single handedly to the elaborate reproductive process. Surely it must be too much investment for little birds and involves a biological economy of choice between singing and breeding. A bird that is invested in singing can do so only at the sacrifice of his reproductive privilege. There is no doubt that singing and breeding are related activities and perhaps stem from the same biological protocol. But little birds cannot do both at the same time. It is reasonable to argue that birds that produce elaborate and sustained singing all their life time contribute minimally to the reproduction and very often are not the breeding males and vice versa.

To hypothesize that the singing bird is not the breeding bird also supports the concept of species as a gene pool. This proposition advocated by Theodosius Dobzhansky in the 1940's was derived from Darwin's second principle of evolution that holds that there is in nature to maximize the genetic possibility of a species stock. Species are bound to exist in their as many possible gene combinations (alleles). So it seems appropriate to expect that the singing bird is not the breeding bird.This can also explain why many bird species that are noted for songs show song variations between members, some of them having greater capacity and some less. Even the song notes may be different among members of the same species flock living in a territory. How do these roles get distributed among the members like they do? What is the resolution that takes place? Yet in this resolution neither the singer nor breeder is the looser. Each participates in a reproductive praxis that fetches him a respectable place in the flock.

Let us turn to the second assumption about bird songs: It is claimed that they play a significant role in mate selection a good singer always is a better bet as mate. The latest version of this theory is that of songs and displays as mechanisms of SMRS or specific mate recognition system. Now this conclusion is derived I believe from Darwin's fourth law of sexual selection stated in his second book published in 1871 The Descent of Man and Selection in Relation to the Sexes, and its revision in 1872 The Expression of Emotions in Man and Animals. It is thought that Darwin was inspired to formulate his law by the peacock and his extravagant tail display which he concluded exerted an influence on the female in her choice of mate. The elaborate tail and its display had reproductive advantage for the male bird. This was the only way he could understand this extravagant but useless adaption of the creature. Experiments since then and in contemporary times conducted on all kinds of animals with such elaborate but useless adaptations, such as cockscomb, crests and tails it is thought have proven this law (Alcock,2005; Burley, 2006)). This same principle has been extended to understanding bird song.

Actually this is persuasive in itself but for the fact that empirical observations in the natural setting contradict it totally. It is amazing that no one has ever cared to do a field study of peacocks. Had they done that they would have discovered that the peacock is not dimorphic as is the popular belief (and perhaps they are bisexuals, polarization emerging in pairing stage). Both the members in the breeding pair have elaborate tails (tail coverts mainly; the actual tail lies underneath the flamboyant covert). During the breeding season starting June-July the breeding pairs start molting and by November have completely lost the coverts. This loss is permanent;

Pictures 5 & 6: Peacock (breeding member) in June and (displaying member) in November

they do not regain the colourful feathers. In fact this molting is a communal event, since these birds exist in large flocks of 20 or more during these months when all the breeding members in the flock lose their coverts en masse. But, corroborating my theory of a purely displaying role in the flock, there is always a member (perhaps a pair) in the flock that does not lose the coverts. Another factor is that during this time a maximum amount of vocalizations can be heard, possibly produced by the breeding members suggesting role distribution. Display of feathers can be had only in November post breeding I believe when the chicks are already out fully fledged. This

Picture 7: Peacock breeding member in November with complete loss of tail coverts mistaken to be female

suggests the possibility that displays are directed at the newly fledged chicks rather than the mate. In fact a free ranging study of peacocks in Japan (Takahashi *et al.*, 2007) has opened up the arena for further speculations on the truth of the display-sexual selection nexus.

Similarly field observations on birds like orioles, barbets, iora, robin so on indicate that the song bird achieves his peak capacity for song only in post hatching even though he keeps this capacity till the end of the season. In many of these instances he abstains from mating completely in order to achieve this even. There the question of role of song in actual mate selection by the female does not arise at all. In addition among many of these birds it is the female's vocalization that initiates the season. Among some others like leaf bird, oriole and bulbul she produces visual displays such as rolling while the male has not even inaugurated his song. Such contrary empirical evidences point out that the purpose of song must surely be quite a different one.

It must also be pointed out that among most birds under observation here the question of mate selection via courtship seems to be a marginal one. Konrad Lorenz's (1949) concept of imprinting might better explain pairing among birds. Since among most birds the breeding pairs come out of the same season's clutch and among many like iora, sunbird, drongos, leaf birds, tree pie (pallida), oriole, tailor bird, bulbuls, thrush and barbets the eggs are laid in the same nest. Among Tickell's flycatcher, robins, babblers, malakoha, the mates belong to different seasons but still the process of imprinting can explain pairing, for in each case the male bird is present in the female's immediate environment from day one. Imprinting means the pair recognition is achieved at the early critical period of six to seven hours of hatching or probably even in the egg stage. This means pairs are formed long before the songs are even learnt leave alone first song production. Among all birds song capacity is not reached until fully fledged and among many others its full performance may come even later after entering breeding as I said above.

Among the various type of surrogating arrangements such as Tickell's and magpie robin the imprinting mechanism can be very complicated. For instance a Tickell's prospective breeding female is adopted in her just fledged state and continuously reproductive fed by the unpaired male (younger of a twin) who seeks a mate from her (but not to mate her). During this time they are together and freely intermingle until she reaches sexual maturity in the course of contact with him. Subsequently she consents to mate with her true mate (the older twin) by sitting on top of him. She continues to intermingle until she produces eggs. Similarly the magpie robin surrogating male is allowed free contact with his female twin (they grow up together in the same nest and later uptill breeding season for nearly nine months between fledging and first breeding)) even if she is betrothed to a male from an alien clutch/territorial location. It is her twin that mingles with her freely until she attains sexual maturity and successfully mates with the older breeding male. In both these cases it may be possible that the presence of two males is interpreted as the need for sexual selection on the part of the female and that she chooses in favour of the better breeding male. It is needless to remind here that this breeding bird of her choice is not the singing male in the flock. In fact among magpie robins he has stopped producing song long before and it is her twin's song she relies on to safeguard nesting territory. Moreover the breeding bird is alien in both the cases either by age or by allopatry and it is my suggestion that the females in both the cases make an altruistic choice to reproduce for the survival of the clan or flock by giving up their primary objects of love (see below for more on this).

What could have led to the adaptation of such a mechanism of song and other displays among birds? What is the nature of song capacity? What was the nature of problematic it resolved? It may be noted that sexual selection theory that views song as part of Specific Mate Recognition System assumes universally that such display impacts on the female bird and that too only in a certain predetermined way, the way of mate selection. It also assumes universally that it is the female who does the selecting (this belief has found currency as Darwinian feminism). It also assumes upon sexual polarization as pre-requisite for breeding. It is clear this understanding of song views them as mechanisms for stronger pairing for the purpose of breeding, sexual polarization being the other important adaptation. By sexual polarization is meant a receptive pliant timid maternalized female and a dominating aggressive strongly territorial male.

It seems to me that the reliance on complete sexual polarity as desirable to reproductive success is an extremely dubious one. Because a totally weakened female would mean she is not fit for the breeding. Further a completely polarized male bird can easily desert the totally vulnerable brooding female and her chicks. In such instance it would require the female to acquire displays of all kinds to hold the male back. In view of this I suggest that the sexual selection theory does not formulate the nature of problematic sufficiently by reducing it to appropriate mate selection. In the natural state even in the state of war mating would be the least threatened because mating requires aggression. The problematic I suggest should lie in the post mating stage in ensuring the reproductive success.

I suggest that displays are psycho biological mechanism acquired by birds not to strengthen bi parenting pair breeding but to counter its drawbacks; one of it being the rise of complete sexual polarization among breeding pairs and a second being the isolation of breeding pairs from its flock possibly an effect of the first adaptation. Somewhere in the process of evolution birds must have realized that the adaptation of sexual polarization (bi parenting) was detrimental to the reproductive success. Sexual polarization was found to disrupt the completion of the reproductive process that followed; because reproduction required both the parents to be maternalized. Foremost song capacity is nature's way of circumventing the possibility of complete sexual polarization in the breeding birds in order to ensure successful completion of the reproductive process. That's why in most cases (including pea fowls) the bird that displays only initiates the breeding process rather than take on the actual role of

Picture 8: Spotted Owlet

the breeding mate. And his continued display further ensures the completion of that process by suppressing unwanted sexual polarization. This explains why most breeding males contribute as much to the reproductive process as the female and are as maternalized (unlike of course among humans).

Secondly songs produce group cohesion and herd specific flocks within a population to focus on the reproductive activity. It loosens the pair bonding with a view to forming coalitions that involves the flock. Birds that sing and produce displays are not breeding members but monitoring birds. They rule by charisma and leadership to herd the flock together towards a more altruistic activity of breeding. Songs of birds are the manifestation of herding instinct. Songs and displays are acknowledgement of need for social co-operation among a community of animals. Thus songs of birds are social in nature. This social cohesion is not different in kind from colony formations among animals placed in the lower orders; the difference is only in degree. At the most fundamental level it enabled to strengthen species consciousness among the higher order individualized birds. Further it enabled awareness of kinship among fellow species. Songs should be viewed not as a mechanism of specific mate recognition system but as mechanism for recognizing specific mating systems. Songs remind a bird that they belong that flock, species identity. Songs should be viewed as aspect of a reproductive praxis rather than individual female mate recognition. Songs not only organize specific mating systems or reproductive arrangements but also their relation with other reproductive arrangements in the local community.

There is much empirical evidence for this even among diverse species such as frogs, cats, bees and insects, *etc.* The contemporary bias on breeding display including bird song as sexually selected (see Podos *et al.*, 2004 for an account) seems to be based on insufficient observations and quantitative approaches. For instance among frogs it is thought that chucks emitted by the members in a population carry significant information to the female to make a choice of mate. The selected signal is of course the biggest chuckle emited by the biggest frog in the population (see Searcy & Andersson, 1986). The truth is among frog species the male actually mates with a smaller sized surrogate male and fills his penile pouch with sperms which is then transfered to the receptive female sitting in a suitable environment (animals do not have a ready made penis but merely a hymen, which can be projected in or out). This is because the actual breeding male is too big size to sit on the female for as long as the fertilization is taking place. Since this process takes place outside the body it involves a long period of time. The chucks appear to be more likely therefore species signals for recognition of specific mating system directed at the flock rather than for mate recognition. Similarly, Experiments that seek to prove the song as sexually selected trait are all conducted on individual members (and their responses to stuffed dummies with speakers) rather than the mating system. Members, particularly the females in their post breeding season are selected for the purpose. Such experiemnts place their conclusions on merely poses struck by the estradiol implanted females in response to playbacks of recorded species song emanating from the stuffed dummies. It is thought that certain poses, such as beak and tail up with wings away from body, indicates sexual solicitation gesture (see for instance Searcy, 1989). My field observations suggest that this gesture is typical among the avis as a class and indicates paternity or maternity. This display is usually produced in the post breeding once the chicks are out. This is also the time when the greatest quantity of songs is produeced. In fact it is possible to reinterpret Searcy's data that breeding song of birds are directed at the fledgelings and are usually produced during post hatching stage as my field data indicates. It would appear that sexual selction theory is based on mere communication gap between humans and birds!

Let us now proceed to construct a possible scenario to understand why there is a lead singer and why he is as a rule male. We know that neo avis speciated from primitive reptilian birds. This first generation neo avis must have faced a crisis when laisse faire biparenting lead to their extinction (K-T extinction boundary). Subsequent generations developed displays and vocalizations to distinguish between critical breeding populations and to maintain its cohesion. This must also explain the development of birds from precocial to altrical. By the by elaborate songs were developed by younger species. Display and especially song must be a part of the collective memory of avis as a class and of every bird species rather than individually held. That is why the species typical song is not learnt but a fledgling merely recollects it by recognizing it. Every generation of lead singer in a bird population plays the seminal role of actualizing the collective memory, and redeploys the song mechanism for species cohesion and perpetuation. Song matching and grammar of avian vocalization must be the measure of the stakes of each species in this collective memory. If we consider displays as a universal phenomenon adapted among majority of the species perhaps we can extend a homological origin for all the organisms. This paradigmatic narrative must underlie the development of communication universally among organisms.

The question of course arises if song is collective memory how did the individualized role of the lead singer develop? There may not be any one story that explains it. If we look at the field data on the role of the lead singer among the various species what hits us is the diversity of song routines as much as songs. It becomes clear that even though the underlying rationale for the song mechanism is the same each bird species produces distinctive dramas of song mechanism.

My hunch is that along the evolutionary line ever too often a young bird must have become anxious for the successful fledging of his mate when faced with continued sexual aggression between the paired members of his community. It must have been a male because nature at the outset had made female strong either ways and for her the problem never arose. He couldn't have took over the fledging nor fended against the transgressors. The only way for him was to herd back the breeding pairs to the fledging nest. To his delight he achieved this and became anxious to keep his song. There was born our song bird. He was male and had barely reached breeding capacity. The choice was drastic because subversively he found that he had to forgo his reproductive role to keep on singing. It required a sacrifice on his part. But it suited him as well as other members because it lead to a reproductive coalition, rather than loosen his role in the flock's reproductive praxis. That's how the leading singer found his vocation; it was his song that kept the birds together. Songs by the by became key to the group/ social dynamics of the entire community. We find several variations of this archetype among the birds of Bhadra that corroborate this scenario *viz.,*: the Tickell's flycatcher unpaired male who puts in his lot with a breeding twin and his female for the sake of having a mate; the iora singing male who herds his parents to complete the fledging of his female born in the same season's clutch as he but a month after him; the Magpie robin breeding male who inducts a surrogating male in order to have a female,the babbler male who delays his own sexual role and plays surrogate to an older male in his flock, even the female lead singer of the tailor bird fits this picture without exception.

My claim that songs are mechanisms of herding and social cohesion and keeps a flock together actually finds a rationale in the literature on speciation. Is it possible to recognize organisms in nature in terms of groups? If so what should be the criterion? In the literature the idea of cohesion is very central to the concept of biological group or in other words species. It is thought that a species maintains its uniqueness as a group or cohesion primarily by the exercise of certain "isolating mechanisms," which cause reproductive isolation (other mechanisms of cohesion are thought to be various types of ecological selection). In fact song of birds are viewed as a type of isolating mechanism that can be termed as behavioural (Mallet in Calow, 1998), which bring about cohesion among a local population for the sake of reproduction

The term isolating mechanism was adopted by Dobzhansky in the 1930's and then later given currency by Mayr in 1970's. Dobzansky viewed song of the breeding male as isolating the breeding female from other prospective males of same species as well as other species (pre mating isolating mechanism, Mallet in Calow, 1998). And therefore it produced species cohesion even though he viewed it solely from the perspective of the mating pairs. But more recently there has been a trend to reject this concept as bringing about species cohesion, especially in the observations of H.E.H. Paterson (Mallet, 2001, Paterson, 1980 in Levin S *et al.*, (Eds.) (2001, 2007)). He proposes pre zygotic sexual signaling and specific mate recognition system, *i.e.*, displays as intimate communications between the specifically mating pairs (something like love talk). Deriving from his suggestions songs of birds are assigned the status of SMRS in current research. Paterson goes to the extent of rejecting the idea of a cohesive species and suggests that there exists only species recognition. In his view SMRS need not entail even species recognition but only female sexual selection or mate choice. Apparently species cohesion was a spin off come by genetically optimized individual female sexual choice (the female chose the best suitor). Even though this theory appears to be attractive one must be cautious about its ideological underpinning.

Underlying Paterson's framework is a very deep rooted notion of (socio-biological) atomism extrapolated from the apparent phenomenon of individual ontogeny of organisms. In addition it also assumes sexual laissez faire individualism. This atomism and laissez faire is offset by a third notion that of genetic pre-determinism. First of all, in nature organisms are never born individually and it is a rule among birds mating pairs are already paired off at the stage of fledging and as I say above the concept of imprinting might better explain pairing. Secondly such pairs are born into an already in place reproductive praxis; I would say they find themselves in media res (in the middle of things) as far as the reproductive process is concerned. Any framework of understanding of song should therefore incorporate the idea of a group dynamics at least at the level of a minimal flock or a local population.

Paterson's framework assumes that displays are produced and exchanged between breeding members, in most cases even in polygamous animals pairs, which is not the case as I gathered in my field observation. Even if one defines songs as a type of sexual signaling it does not function between breeding pairs but amongst a flock of like members which includes non breeding. The bird that gives the sexual signaling is as I noted not the breeding one. In all the cases for which I present field data the singing bird is reproductively dysfunctional in one way or the other and contributes inversely to his display capacity. There is also the other side of it what happens to the song of the other members, especially the breeding member? In fact looked at as sexual signaling it would seem more appropriate to interpret songs and other displays as negative signaling that impart the reproductive dysfunctionality of the bird, something parallel to the function assigned to castration complex among humans.

Paterson's idea of SMRS has also provided a framework for understanding song type matching found among birds in a local population. This framework foregrounds rivalry or competition over mate choice. It is true that empirically it appears that greater the genetic nearness greater is the incidence of song type matching interpreted as rivalry over mate and niche. We find more song type matching among related birds, especially subspecies members. But there is another way of understanding this which Paterson overlooks. That song matching is the manifestation of kinship structure between the diverse species. Song matching do not reflect rivalry but on the contrary the kinship affiliations and as well as exclusion. It intergrates a species at the level of not merely the local flock but subspecies and clade. Songs involve a social dynamic and not sexual.

It is clear that we need an understanding of the song mechanism as a far more socially dynamic phenomenon than sexual signaling for mate choice. And I suggest that a better concept lies in the ideas of kin recognition and kin selection. These related ideas were proposed by W.D. Hamilton in connection with eusocial insects in 1964. According to the idea kin selection a member of a species can take an altruistic decision to collaborate with his flock even if the decision results in a disadvantage for him especially reproductive disadvantage. According to Hamilton there is a higher likelihood for a member to adopt members of his own genetic make up as beneficiaries of such altruism on his part called as kin recognition. These mechanisms are justified by the selfish/sentimental gene that urges members to take decisions at group level directed to survival of a specific genetic make up rather than mere individuals. Hamilton also proposed that kin recognition and selection are social cue based rather than merely genetic protocol. Song must be such a social cue for genetic cohesion.

Actually there has been a long debate since Hamilton in the literature that opposes kin selection to group selection (Maynard Smith, 1964; Wilson, 2005). It has been also argued that group selection contradicts natural selection (Williams, 1966). This debate stemmed from Hamilton's own research that extended kin selection to group selection and carried on colony insects and haplodiploidy among organisms such as bees (Wilson, 1971; 1975; Brown, 1974; Trivers & Hare, 1976). Hamilton also developed the concept in the form of theory of inclusive fitness at the level of group (Hamilton, 1964; 1976). But subsequent field research has lead to disassociate group selection from kinship and promote the opposite *i.e.,* kinship is a dissolving force in groups (Trivers, 2002; Wilson, 2005). In this regard there have been also reformulations of Hamilton's rule for kin selection to incorporate competition from other related members (Queller, 1992) in the group.

Going by these altercations one would think that kin selection is a fairly debunked idea, since the notion of gene cannot be actually viewed from a merely individual organism but only a critical population: So that there cannot be kin selection without group selection. But it would appear that these controversies rest on improper understanding of the nature of the gene and failure to distinguish the deep mechanisms of gene expressions from purely social. Even in colony formation kin selection should have a prerogative over social selection. Moreover Hamilton formulated his concept in connection with reproductive altruism. In fact regarding Hamiltons' rule it is suspiciously naïve as to what is meant by genetic distance as it is calculated currently. The Hamilton calculation is based on the rule that each parent contributes one half of the genetic material. Total exogamy is the guiding force here. On the contrary I would point out that mating among species is governed by degrees of genetic distance (or nearness) fixed appropriately very likely sequentially in successive mating in a critical population (an endogamic community). So that greater degree of genetic relation (nearness) between parents would give rise to more alleles (maximize offsprings with same genes). It appears that among bird species this genetic relation is of a different degree from humans. This begs to reexamine the nature of genetic transfer by sexual reproduction. Trivers (2002) for instance concluded against kin selection in colony formation on the basis of parent offspring studies. It may be pointed out that the genetic relation between parent offspring may be counted as furthest if we see genetic transfer as a vectoral force which cannot retrace (especially among humans). Futher, it should be kept in mind that within a gene pool, our closest kin - say two sisters - would be genetically furthest between them they being near perfect allelles of each other. On the other hand non alleles of a population (more distant relatives) would be more closely related genetically. Those members totally unrelated or of other species would be furthest. Instances of kin dissolution in insect colonies reported by Hamilton (1964) and later investigated by Trivers (2002) could be merely resizing of the kin based colony to critical level so as to stem the dilution of stock (regression) caused by the increase in breeding members above a certain (species specific) number. This seems to be the secret of a Mendelian critical population.

It would appear that the nature of Mendelian population has not been understood correctly (see later for more on this). Songs of birds can be viewed as social cues for such a herding or cohesion primarily for the sake of reproduction but above pair breeding among genetically related species. Thus also song learning and song matching phenomenon among genetically related species as well as in a local population; not only that the above preposition can better explain the unequal distribution of song capacity among fellow members of a flock. That is it throws interesting light on why it is that there is always only one lead singer in a flock. And why he is not the breeding bird.

A word must be said about the female's reproductive role and kin selection. I have already suggested in the previous pages that the females among the flocks that involve a surrogating male make an altruistic choice to reproduce for the survival of the clan or flock by giving up their primary objects of love. Among the Tickell's the female's choice can be viewed in the light of kin selection and inclusive fitness where she maximizes her reproductive output so that all the males in her flock are provided with a mate. She as we know produces three eggs in her first season a phenomenal out put which loses her of colouration, the species song and then in subsequent seasons complete loss of capacity for song. Similarly the magpie robin female gives up her primary object of love to maximize her flock's reproductive chances a sacrifice equal to that of her male twin. Among the ioras the female of the singing bird commits suicide by consuming poisonous worms when the next generation breeding female has fledged successfully under her care and is ready to take over the breeding role for the flock.

Pictures 9a & b: Asian Paradise Flycatcher female (pre and post breeding plumage)

The female sexual role in nature at the outset is an altruistic one because potentially a female can always take on a male role but she forfeits it for the successful completion of the reproductive process. Take the example of female of the paradise flycatcher, and racket tailed drongo, they take on reproductive roles even at the cost of permanent loss of body parts such as their tail streamers and so on. In addition field observation also reveals that bird species enforce regulation of sexuality of its flock members, which may have an impact on sexual difference. This regulation may have several forces acting behind it: for one it appears that maintaining a critical level population can optimize the claim over niche within and between species. It is also possible it also enforces species cohesion. Many species take recourse to suicide even to maintain this critical level. For instance among iora and Tickell's older birds may simply give up their life by consuming food that causes extreme irreversible molting that leads to a quick death. Among barbets and ioras the population is regulated at such a strict mathematical calculation even at the destruction of an egg or hatchling so as to avoid crossbreeding. Sometimes preying birds may be inducted to carry out the chore such as among leaf birds.

Allopatry is a chief regulating mechanism. It involves ecological selection and can function in more than one way. It can be used to regulate the sexuality of members, one way to keep the population down. We already know the practice of allopatric maintenance of two different population body types among magpie robin and tree pie (parvula) for more efficient breeding. Among these species the same parents can raise different population for interbreeding. Even among ioras the intergenerational differences in song capacity and breeding capacity seem to be partly due to adoption of different ground types for breeding. It is possible birds like the Tickell's flycatcher and parvula tree pies use differentiation in ground type to regulate the sexuality of the offspring and also the reproductive capacity in tune to the need for either sex for mating. The uncontested endurance of such practices in nature strongly support the kin altruism theory.

Let me discuss one more very seminal field example of breeding time display as kin group selected and as a herding mechanism. This example is important because it forms the basis for the formulation of one of the most flambouyant theories of breeding displays, that of Amotz Zahavi the handicap principle (1997). Another related concept is that of Marelene Zuk (1990, 2001), bright mate hypothesis. Both these formulations come about post Paterson and are constructed within a general theory of animal communication called as honest signaling. Organisms are said to produce a range of displays from honest signaling to illegitimate signaling depending on whether the addressee is the real mate or merely an illegitimate intruder. An honest signal is defined as that display which conveys truthfully the inner fitness of the displaying member. The argument is that extravagant breeding displays that are actually handicaps are used as marker or proof of the member having coped with it and of having still remained capable. Even in their extravagance they are honest signals of the member's capacity and therefore such displays play a significant role in female mate choice. Zahavi's study includes a wide range of displays and does not distinguish songs as in any way different or give a theory of song. His conclusion was primarily based on field study of Arabian babbler, but also included primate displays. But contrary to Zahavi's formulation of a Samson-like supermale, going by my own field observations I suggest that even if the displays of the babblers signal a handicap, the handicap is indeed undoubtedly real. The displays are kept at the cost of reproductive capacity. So that in a breeding flock the actual breeding male does not produce any display but this role is carried on by the surrogating member. It is not true that the display is co adapted to a viability, but rather it seems to be inversely adapted to breeding. This elicits a theory of breeding displays as kin (group) selected rather than for mate choice. In the following I present my own field data on babbler species found in the bhadra region as supplement to that I provided in the section in chapter Two.

Babblers such as common and jungle are born in a good size clutch of either five or seven members. In such a clutch there is usually only one male the rest are females (or excluding one other member who is female all others are neutered). When this clutch enters breeding season the breeding male always comes from a previous clutch and is older. The younger male in the new clutch seems to be superceded by this older member and does not have claims over any of the females that were born along with him. But still during the breeding time he remains with the flock and plays what I call as a surrogating role mainly by his ability to produce displays. The breeding male does not produce any vocalizations or physical displays such as retracting the neck so as to make the feathers stand out and chin wagging. But the bulla is used to carry stones to grind fresh wholesome food during the breeding time for the incubating female and the newly hatched chicks. Among common babblers the bulla is also used to carry wet clay to build wattle incubaters. These incubators are built at the base of trees on the ground. It is the younger eclipsed male that can produce the displays but it appears that this is produced at the cost of use of the bulla for feeding purpose. Instead during breeding time this younger male is found leading the flock for foraging; vocalizations and neck displays are produced during this time and may be especially directed at intruders to their territory. It is possible that the breeding male can produce displays but it would not be cost effective for both the males to play the same roles. Since the younger bird cannot yet use his bulla as a storing device.

Considering the role of the displaying bird it elicits the conclusion that displays among babblers cannot be for mate choice but a selection for kin. There is no doubt that in the breeding coalition the displaying bird plays as important role as the dominant breeding male; safeguarding foraging territory. In addition the female who actually comes from the same clutch as the displaying bird seems to be co adapted not for display because her mate choice is determined by the capacity for efficient reproductive feeding. It is possible this is explained by imprinting she undergoes under the reproductive caring of her male parent. The kin selected display is altruistic in nature because the surrogating bird has selected to delay his reproductive advantage in favour of the kin group, by which he is rewarded of course when he gets to pick a female from his elder kin's successful brood.

At the out set Zahavi assumes pair breeding and does not distinguish that breeding birds engage in a reproductive coalition which involves more than the actual mating pair, with members playing different type of roles in the junta. It is my suggestion that what is identified as handicap signaling is an aspect of primarily niche sharing behaviour for optimizing breeding success and only marginally sexual mate selection. Further such a display may not be species typical but is role typical and is exhibited across species, even though such surrogates do produce species typical displays all the time. For instance among the brahminy kites a male member born out of same clutch as a female, but allopatrically bred or translocated after an initial period uses up his cache of food resources (rodents, mainly rats) at a very fast rate and then shacks up with the female during the incubation and hatching stages of her nest of eggs. He does not make up her first mate (*i.e.*, the eggs are not his, he remains reproductively inviable at this stage) but for the better part takes the role of a surrogate parent, while the actual mate never lives in the same nest as her (she could have more than one such conjugating mate). His surrogate parenting role involves babysitting for the female while she ventures out in search of specialist food for her and perhaps the chicks. The surrogating mate is allowed to use up the food cache (rats and other rodents) around the nest. It is my suggestion that adult brahminy kites cannot survive on rodents after a certain age and do not gain reproductive capacity. They usually are found to feed on offal and meat of larger animals. The ability of the member to continue to survive on rats and prolong his childhood turns out to be advantageous during breeding time because he can still survive on food found close to the nesting eco habitat and contributes to the fledging. Such surrogating members can produce displays that imitate younglings display (*e.g.*, paroxysms and fit displays such as puffing the neck feathers and shaking the head or crouching) and to parents. Such typical displays are better labeled as role typical and may be found in unrelated species that exhibit similar niche sharing among fellow members such as cats, frogs or toads for example. Among cats for another instance it appears that the presence of a younger male kin prolongs the rut of the older male and increases successful mating even to a greater extent than if the younger male were to mate his own female (born in same natal litter). While on the other hand the presence of the older male delays maturity in the younger so as to increase the flock's chance of having a breeding male all the time. The confrontational displays exhibited by the males of this species are comparable to courtship displays between the actual breeding pair and can continue to occur even when the female is removed from the venue. What we have here is a specific type of mating system rather than rivalry over mate choice *etc.* Thus species form coalitions for breeding involving more than the mating pair and where roles are distributed among the participants. It involves reciprocal kin selection that is mutually rewarding even though one of the members remains reproductively inviable for as long as the relation is carrying on. In addition when he undertakes to mate viably it is more likely that he selects a younger female (among common babblers the surrogating male sibling gets to pick from the newly fledged clutch).

There is more evidence for displays as adaptations for group selection than sexual selection. Mutually beneficial niche sharing is a very important to the successful completion of the reproductive process and points to the inefficacy of pair breeding in the matter. Not only that such niche sharing also regulates the sexuality of the members so as to give every member a chance to breed. Because strictly speaking this member who surrogates could easily play the breeding role but he chooses to delay so as to optimize reproductive success of his female kin and perhaps his own by delaying it. The appropriate regulation of sexuality among members that share a niche is directed at optimizing a choice of mate (maximizing reproductive success) rather than at optimum mate choice (the reason for this is discussed later). Zahavi remains a staunch critic of kin selection theories (1995) but it must be remembered that reproduction is at the outset an activity involving altruism; for its gains are beyond the individual members.

Zuk's (1990) study of female mate choice among red jungle fowl also concluded the preference for males with larger combs and other display features. Several loop holes are evident in Zuk's experiment. My own field observation reveals that display plumage – crest, wattles, long tail feathers and bright plumage colours- develop on a singular member during the post mating-hatching stage in its flock. The comb is smaller initially but develops to full size as the season proceeds, similarly plumage colours. Sexual assignation to this individual is made uncertain by several observations. This member is found to be in charge of the fledging (can be found leading the chicks while foraging) and plays a female role.Furthermore post development of the display features, the member produced fewer eggs (a single egg) in subsequent seasons accompanied with lot of vocalization, typically assigned to cocks. Such afully developed displaying member is completely neutered and reproductively dysfunctional in every way. I conclude that among the jungle fowl species the leading reproducing female acquires the display features but is neutered in the course. Other members in the flock who remain plain are surrogating birds who are in charge of the hatching but also behave as a reserve breeding force in case of any ecological pressure on the flock.The display appendages, especially the long tail feathers are acquired by the leading female to monitor the chicks and function in the same way as the tail of any other animal such as that of a cat or a dog. The tip of the tail is used to sense young

Picture 10: Red (domesticated) Jungle Fowl in -breeding first season plumage fledging the chicks

litter. The species are otherwise not sexually dimorphed at all. Any differences in plumage colour and size of crest among members of a flock is therefore merely indication of different stages of breeding in the flock rather than sexual markers. Secondly, it is also possible that they are (and perhaps all species classed under fowls) bisexual or perhaps asexual. Until sufficient observation is done on the breeding roles in the flock I suggest Zuk's hypothesis should remain inconclusive.

Zuk's theory of Bright mate is derived from the research on carotenoid-rich food preferences among birds and other animals. In birds carotenoids enhance the plumage colours. It is hypothesized that selection for the bright plumage must play a role in mate choice by females. But there is a catch here. Carotenoids are known to cause fitness loss. So it is concluded that the brighter mate is the one who has survived the selection for the handicap. But the use of carotenoids to organisms is not so straightforward as it is so made out. Organisms that reproduce sexually are, throughout their fertile years, biologically subjected to periodic sexual cycles that have two phases. They undergo a waxing stage when the breeding capacity is slowly enhanced and peaks; the other is the waning when the mature ova or sperm is ejected either way even if fertilized. The two cycles require two different sets of hormones to be effective. The production of these hormones in the body requires a fat rich diet. Some of the most efficient natural sources of such a diet contain a high degree of carotenoids. In fact carotenoids are nothing but polystereates or fatty acids (lipids) and enable the pituitary to produce hormones required for the the sexual cycles in organism. Even though carotenoids themselves are not harmful the increase of sex hormones can cause lowering of the body's immunity to parasites and sickness-like condition or fitness loss. Infact in a recent experiment (Hamilton & Zuk, 1982) it has been shown bright plumage is related to the parasite load. In birds during the waning stage the sex hormones produce molting, an extremely discomfiting condition in a young and healthy bird but can cause even death in older members. But carotenoids also cause brightness of pigmentation of the organism as a spin off. Consequently bright plumage is correlated to the reproductive capacity but inversely so; for brighter plumage would mean that the organism is undergoing a waning cycle and unavailable for mating. In younger organism the cycles may be shorter and the inviability more persistence, this would remain until it has reached certain mature growth. Thus younger organisms can appear brighter always. But it is clear that colourful displays (including song) are sexually correlated features almost universally.

It is important to see that carotenoids are unavoidable to the healthful completion of the sexual cycle in both sexes and are a universal aspect of the biology of all sexually reproducing organisms. Given this kind of sexual biology organisms can always potentially face with inviability. In fact greater the fitness of an individual in the first phase greater is the inviability in the second even if this means brighter the mate. If entering into real breeding the greater fitness can lead to greater risk of miscarriage or death of parent during pregnancy or impotency (large sized sperms that produce inviable zygotes). In fact the best solution is individuals can opt to be less fit so as to minimize the risks because the less fitness of the parent may not affect the zygote seriously so as to not be healed in the post natal stage nurturing. Thus in reality the brighter mate need not be the best mate biologically. This is of course corroborated in my own field observations. The breeding birds are always dull and without displays or song.

Among many birds an option is made for low fat or fibre rcih diet. But this usually leads to dull plumage. Among racket tailed drongo the main diet is composed of flies and other insects with glassy wings, which means low fat accompanied by high methyl acetate so that the birds have developed glossy black plumage and racket and tail streamers. And during post breeding the breeding members of this species cannot help the loss of these appendages. Among other species such as iora and Tickell's flycatcher a compromise seems to have been struck. That means they have selected for brighter plumage and a balanced diet. But the hitch is during breeding the breeding members lose their bright plumage. The iora breeding male turns black his female molts painfully. In the Tickell's flycatcher, the male is found molting and his female becomes paler blue during post breeding. Yet it may be still agrued that the handicap principle is still functional if members appear brightest and still mange to produce viably even at the cost of the display once entering breeding. But this seems not to be the case. It appears that displays play the role of a negative signal in mate choice and seems to be negatively sexually selected in keeping with the inverse relations between display and sexual health.

Let me explain, among the iora at the beginning of the breeding season the breeding male and the singing male show different responses to consuming carotenoid supplying nutrients. The breeding member grows a black mantle on his entire upper plumage and loses all his brightness, while in the singing member the black mantle is limited to the cap only. This ability among them appears to be directly related to the allopatric differential feeding of the chicks practiced by the parents during fledging. Iora diet is mainly composed of loopers or caterpillars of geometridea moths during the rainy months or their grubs in the dry months. Caterpillar of the *Selenia tetralunaria* species (purple thorn) is especially favoured for it feeds on colourful flowering trees like the red asoka and flame of the forest and is rich in carotene. But they are found aplenty only in the rainy months. In winter the diet is mainly composed of moth grubs of geometeridea such as chiasma species. The availability of these nourishments coincides with the fledging stage of the two annual breeding events. The member destined to be the singer fledges in May –June when the moth loopers fattened on pollen are found aplenty in the environment. The members chosen for breeding are fed with low lipid moth grubs. The breeding male even though fledged like the singing male during the rainy months still gets to feed on less colourful caterpillars (other than purple thorn moth) perhaps because he has to share the niche with the older singing male. The female chick feeds on moth grubs (moth caterpillars) only since she fledges during winter time (November –December) when grubs are found maximally. This is not all, it appears that the carotenoid rich infant diet does not merely enhance the feathers but enables early song capacity in the singing member. In the singing member this also leads to complete reproductive fitness loss, while the other is reconciled to differential song capacity.

Hormonal and nuitritional studies of song have proved that corticosterone an adrenal hormone produced by the stimulation of the pituitary by carotene rich fatty diet has a big role to play in song production (Nowicki *et al.*, (1998, 2000). My field observations point to the fact that certain of types of carotene rich foods early in development stage either speedens the song production of the bird; the species typical song is produced at a very early age. In a fully developed member continuation of the same food enhances the song capacity by increasing the repertoire even though the species typical song may be completly replaced by then. At this time what is affected is the nature of the song notes rather than the capacity, infact the quantity increases phenomenally but short frequency notes are produced for a longer time. Usually the differential song capacity of the breeding and singing members appears to be mainly due to allopatric differences practiced by the parents in raising the fledglings, denial of certain type of food to one offspring or over feeding another so on. Corticosterone in the early developmental stage appears to speeden up the song capacity but at the later stage enhances the reproductive capacity. It is also seems logical to expect the same experimentally because we know that most animals practice signaling when under carotene nutrition induced stress, for example song matching or structural mimicry. It is not possible that organisms have adapted for such a defense mechanism if stress induced hormones such as corticosterone reduced their signaling capacity. On the contrary it must enhance it.

As I stated at the beginning of this discussion Zahavi's Handicap signaling and Zuk's Bright mate hypotheses are constructed within the frame of honest signaling theory of animal communication. Central to this frame is a fascinating theory of games derived to encapsulate Darwinian maxim survival of the fittest as the basic evolutionary process. Theory of games was first proposed in the context of economic behaviour of humans (Von Nuemann & Morgentern (1944). But sociobiologists soon found it useful to explain evolutionary behavior Smith & Price (1973); Smith (1974, 1979, 2003). The classical game theory model posits two opposing strategies among contestants, labeled as hawks and doves. Hawk strategy is aggressive while dove is that of retreat or defensive. On the evolutionary front it is predicted that in a stable population there are bound to be members adapted to be hawks and others to be doves since it is not cost effective for all to be hawks or all doves. It is predicted that in a stable population there are bound to be more doves and only a few hawks since this strategy is not cost effective. Displays and songs have been viewed as aggressive signaling that fit the game theory (*i.e.*, game of signaling). An honest signal reveals to the opponent the cost involved in fighting and therefore the stronger signal will deter the rival from taking on the hawk strategy and to opt for dove and retreat (see van Staaden *et al.*, (2011) for an account).

The game theory can be actually formulated in two modes confrontational, as has been done for understanding sexual signaling for instance, and the other non confrontational, as in ecosystem analysis (Smith, 1976) for instance, even though in contemporary ornithological literature there is a preference for the first. Real life instances such as claims over a common food resources, territory *etc.* are usually understood in the confrontational mode. It must also be pointed out that both these scenarios are constructed on a one-to-one relation among members, *i.e.,* the game is played between individuals. And the adaptations are also individualist. Moreover in both modeling ratio of fitness cost between individuals is always formulated as inversely related, even though in the second model in a stable population optimum fitness ratio of one is predicted as in frequency selected co adaptations or conditioned responses. This is an inversion of Hamilton's inclusive fitness hypothesis. In fact there is a general opinion that game theory and group selection are mutually exclusive ideas and that the latter contradicts the principle of natural selection which is individualistic (see Alcock (2005) for an account, p. 18). Even so the second mode may be extended to understand group selection as in a team game such as football or cricket. In such group games one will have to account for within group strategy relation as well as between groups. Within the group fitness costs may not be inversley related but the selections may be' as for instance in a capitalist economy the relation between producers and consumers.

Consequently, signaling game theory at the outset assumes signals, especially sexual, as individualistic adaptations. In addition the sexual game is played at the individual level. Also it is predicted that a stronger hawk signal must elicit a dove response. Contrarily it is my suggestion that in an individualistic game potentially every member, when confronted by an absolute hawk, has more chances to opt for adaptation for the hawk strategy (more fitness cost is any way involved so increase one's fitness, *i.e.,* adapt for genes that outrun the other, the rat race) and continue to fight. On the other hand when confronted by a dove the selection would be for dove (fitness cost of confrontation is any way less). In the former case the worst case scenario would be unresolved conflict where the cost is actually paid at the group level (species becomes extinct) rather than the individual because no one gets anything. When individuals faced with hawks genetically opt for dove strategy therefore it is an altruistic decision seeking a resolution of conflict that favours other than self.

I have one more reason for proposing group selection and for arguing the inadequacy of the evolutionary game mathematical modeling whether confrontational or non confrontational. Considering that the evolutionary game is played by the Mendelian critical population the mathematical model of game theory is far far inadequate. In a Mendelian population the distribution of hawk-dove strategies must be group selected, in the sense that even if displays are produced by individuals at their own cost it is possessed by the group as a whole. The choice for hawk or dove strategy would be determined by order of age, *i.e.,* one who takes the chance first, usually by being born first, would have more choice. Further down the line the choice would go on becoming smaller and smaller if we consider that a strategy may not be viable by dearth of having reached an optimal number rather than by individual genetic selection. Moreover the nature and role of Mendelian population and the relationship between genotype and phenotype in the evolutionary process have not been understood sufficiently. The distribution of song and breeding capacity among birds (reciprocal kin group altruism as I have argued) that my field data presents seems to foreground the fact that strictly the hawk and dove strategies must stem from the same gene for aggression, and the two strategies are merely expressions of antinomies. Therefore in a population all the members must possess the gene but merely do not express it similarly. That's why a breeding bird does not display and a displaying bird fails to reproduce. The evolutionary game in such a population can be described as a game of kho kho rather than football. In such an evolutionary flow it may be predicted that an allele or genotype selected for the hawk strategy game intitally will have to perforce give it up and move on and would end up as dove. While the genotype selecting for dove initially would stand a more likelihood to inherit the selection for hawk. It is not as if there are so many genotypes correlating to so many phenotypes but there are merely a number of antinomic expressions of a genotype to choose from by the members of a critical population. This game will be framed by the number of turns required among the members to play all the possible antinomies at least once. In such an evolutionary game when individuals confront, the contest is not so much about the choice of hawk above dove but what success rate individuals have with these strategies when they posses them. Ultimately displays of birds raise question as to the very nature of gene as a positive entity as held in science currently. (See my discussion on Mendelian critical population and species evolution below and Appendices 4 & 5; see also my discussion of genetics of Hamilton's rule above.)

In all the discussion above the question remains of course what of the species cohesion I claimed as a feature of the herding mechanism of the song. In classical ornithology species is an evolutionary unit brought about by convergent evolutionary selective adaptation on several individual organism's genes. Since Darwin's formulation of this evolutionary theory in the nineteenth century, several versions of it have developed. Central to all of them is the priority given to genetic adaptations as instigating species differentiation. Once a certain genetic combination comes into existence through processes of natural selection, it may be maintained through various mechanisms primarily by a mating or fertilizing system. There may be differences in the degree of importance given to the concept of mating system to the species coming into existence; as I recounted earlier Dobhzansky and Mayr postulate the existence of isolating mechanisms as a necessary reinforcing condition for speciation, while Paterson denies its primacy. In fact Paterson formulates a stronger species recognition concept by privileging the aspect of genetic adaptation even though he dissolves species cohesion by the feat.

Much of the controversies over the concept of species have been debated around the concept of adaptation and whether to view it as a globally evolutionary or as local event. In its extreme formulation we find in Paterson who suggested local adaptations as secondary and merely stabilizing forces. Secondly genetic adaptation is viewed as a positive phenomenon of the appearance of new genetic structures correlating to morphological and other gene expressions. Thirdly species is viewed as occurring at the level of individual organism. It is my suggestion that there underlies a very basic misunderstanding of what we mean by evolution and selective adaptation, or when we mean that the new gene is better adapted to the (new) ecological situation.

It is my suggestion viewed on my field observations on songs of birds (because songs of birds encode and transmit information on their speciation relation to other birds by syllabic and tonal patterns) that evolutionary process involves not the appearance of new genes specifically adapted but selective loss of "specific" genes. This loss of specificity is not subsequently followed by the appearance of a new specific gene in its place, but merely a suitable restructuring of the depleted genetic materials brought about by adoption of more and more elaborate gene mechanisms/expression, mainly pleiotropy, to maintain that restructuring. Perhaps in the paralance of current genetics the restructuring may be understood as follows: the loci of selective loss of specific gene in its nullified form - since gene loss always expresses as an absence or energy-void with negative value and never an absolute zero (perhaps can be seen as the degree of a gene's loss of specific expressibility) - always becomes the loci for pleiotropic gene mechanisms/ expressions (new adaptations) of the new organism. In fact such pleiotropy may be defined as instaneous momentary restructuring (recombination) of genes suitably to the allopatric context of demand and is the source of mRNA (see p. 137 & 150 for reference), *i.e.,* negative signaling or signaling by absence of specificty. Songs are such pleiotropic mechanisms (see below for more on this).

In the evolutionary process species specifity (and the new adaptations) is a function of the disinherited/nullified gene(s) and results in a fundamental genetic instability. That is to say there is no specificity to hand down the generation but a null gene (complex). This demands for an elaborate reproductive mechanism which can hold the genetic structure togther. Herein lies the origin of the Mendelian population genetics (unlike Dobzhansky who posits a Mendelian population a priorily to the speciation event), which will regulate the proper allotment or apportioning of gene among members and down the generations in a fixed way (*i.e.,* enforce the laws of inheritence) in order to combat this fundamental instability. This is the origin of a biological species, or kin ship structure in other words. The Mendelian genetics ensures not any specific trait transfer (because there is none to do so) but that the species does not collapse into undifferentiated co-membership. It preserves differences between allele members. This brings us to the third objection: it is important to see that species existence (and speciation events) do not take place at the individual level (even paired) but always involve a critical population. This is a logical process because the selection for a particular stock will always preclude the mode of its transfer and perpetuation. The critical population size made up of alleles is important for the perpetuation of that population as a coherent species. A species can exist only in the form of a critical population, which is actually the minimal reproductive unit covering the critical number of generations required for the proper allotment of gene for the species to perpetuate. Beyond this critical population size a species will produce subspecies or population alleles. Population alleles maximize microhabitat differences and adaptations (see Appendix 4 for a detailed discussion). From the point of evolution, perhaps in this event of speciation lies the ultimate cause for the progress of avis class and others from pre-cocial to altrical forms.

My critique of Darwin's origin of species does not discredit the idea of evolutionary change. But I suggest that Darwin hit upon a correct preposition based upon wrong insights. Basic among them is that of sexual selection where he failed to simply recognize the functions of adaptations. For instance he wrongly attributes the male insect's (also crusteacea) prehensile capacity as adaptation for better mate selection while in reality the adaptation is for ensuring successful completion of the reproductive process by securing the spermata inside the ova sac and is present only in the surrogating memeber. But most important among them was the notion of adaptation by natural selection as the underlying process of evolution when in fact the selection for an adaptation does not have a correlation with the selection on genetic restructuring (think of the magpie robin beak structure diversity). Rather both adaptations as well as genetic restructuring are subjected to the rule of diversity for a species/ trait (think of the sunbird beak morphology). He recognized the microprocesses of evolutionary change on mere evidences such as changes in phenotypic morphology of birds and other animals. But the incomplete fossil evidences with baffling historical gaps derailed him completely rooted as his thinking was in Victorian positivism, which lead him to make a direct relation between the adaptations and evolutionary change. I suggest such phenotypic changes are not evolutionary in themselves but creative responses of organic populations to their habitat diversity.

A good proof of my claims I suggest may be found in the display of phenotypic diversity among contemporary members of a species in the same population. Deriving from Darwin's natural selection two theories have been proposed to explain such phenotypic diversity. One is the local adaptation model, a second is disruptive selection. The former promoted by Mayr (1959) seeks a purely allopatric explanation for diversity such as beak or wing size and shape, or claws *etc.*, that members can opt for different phenotypes in order to maximize micro habitat diversity. The latter explains phenotypic diversity in terms of intra species competition and sympatric speciation. Going by the field observations - magpie robin, tree pies, sunbirds, *etc.*- I presented in this book it would appear that these diversity of responses are least to maximize local micro habitat differences (because most species that display such diversity are specialists and all members share the same niche) but more for the maintenance of species cohesion. Such diversity appears to be kin group selected and complimentarily adapted by the members for fitness of the group (example different beak shapes for the purpose of infantile feeding, different beak and claw shapes for tearing and twining nesting materials different shapes and sizes of flight feathers in a flock, *etc.*). It is possible that the legendary Darwin's finches which became the basis for his theory of natural selection were merely an interbreeding population of birds similar to magpie robins I described in this book. This diversity of phenotypes in a local population presents before us the plasticity of the underlying genetic structuring. They must be adaptations made possible by what I have called as speciation by selective loss of specific genes. Only loss of genetic specificity opened up ways for the diversity of responses. Little did Darwin guess that the underlying ultimate mechanism of selection was one of loss of genetic specificity.

In fact the findings of genetic research today fairly corroborate my claim even though scientists are refusing to recognize the nature of the gene. For instance it has been discovered that species specificity is not located on any gene within the cell nuecleus but is revealed by the cytochrome b gene of the mitochondrial DNA that lies outside the nucleus. All members of the avian class must have genetic composition that answers for beak or wing or claw *etc.* The point is how these genes are expressed in specific terms or species specific way of doing things. In fact the plasticity of phenotypes and the trouble species have in maintaining phenotypic status quo in the natural setting throws into question the very idea of genetic selection as the evolutionary process rather it appears to be genetic deselection. Reviewing the situation of beak diversity among the birds for instance, it would appear that beak shape and size develop solely in relation to the beak morphology of the feeding parent and the method of feeding. Chicks usually acquire beak shape and ssize in complimentarity to the feeding parents's. Microhabitat differences in the post development stage merely reinforce the newly acquired appendage. Such is the plasticity of the adaptive process.

This insight into speciation subverts Darwin's evolutionary mapping. His mapping assumes that evolution is a process of more and more genetically specific and complex adaptations. In reality it is a process of genetic dilution and generalization. Higher order (later adapted) animals are less and less genetically adapted and perform more and more functions with less and less genes. Central to evolution is the principle of negation "na iti." Evolutionary mechanism is a continuous process of substraction rather than an appearance or presence of genes That is to say organisms in the universe are in an eternal race against genetic pre-determinism, but the looming chaos or anarchy is held at bay by modes of socio biological organizing.

With this we come to another Darwinian issue, the law of survival of the fittest for natural selection by which Darwin justifies the evolutionary process. For Darwin the evolutionary process was always progressive towards a fitter organism and more developed. The law became the philosophical ground for claim of natural superiority of some species and races over others and grew into a rationale for the supremacy of humans over animals and western over other races. My field study has posed a number of instances such as magpie robin, tree pie, sunbird *etc.* that overshadow the claims of this law. It seems more likely that the evolutionary process is not governed by the law of survival of the fittest but only maximizing the genetic possibilities even at the cost of indivdual fitness; at the level of individual species the evolutionary process maximized the species stock and is kin group selected, homologically on a larger scale genetic diversity is optimized subjected only to the law of the whole. One of the conundrums of evolutionary theory has been why there are no intermediary forms between species in all cases. Darwin comes to the conclusion this must be the work of natural slection and the struggle for existence. It is my suggestion that the Mendelian genetics that I outlined in connection with species existence above must hold homologically at all levels of speciation (or rather all levels of divergence), maximizing the number of allele groups (or the critical number) at every level so as to hold the species apart and preserve their differentiation. It is possible that the absence of intermediary forms merely indicates the different law of inheritance required to maintain this differentiation among allele members at that level, *i.e.*, a purely mathematical condition arising out of the speciation variables and not selection or contest. Law of inheritance is nothing but law of numbers.

Thus speciation is always genetically ad hoc arrangement (because of complexification of gene mechanisms/expression) and is never organized around strict ontological genetic pre-determinism. Molecular research on drosophila has shown that two allele members with same head structures show different genetic make up and vice versa (Carson & Templeton, 1984). But fuzzier the gene and their expressions the more rigid are the biological mechanisms. Many observed phenomenon such as maintenance of a critical population level in a local flock, practice of allopatry for regulation of sexuality of both sexes within the breeding flock and practice of calculated infanticide not merely give the measure of rigidity of a reproductive praxis but points to its ad hoc nature. A strictly Mendelian population is maintained among the birds, which is governed by the mathematical rule of allotment of gene combination among members spread across several generations and over a length of time. But the source of this Mendelian population is not genetic specificity but its instability. The deadlock on interbreeding is strongest because it is an intergenerational one, and is functional as long as the chosen reproductive arrangment is kept in practice. At the centre of a Mendelian reproductive praxis is the missing specific gene. What is handed down by such a Mendelian genetics is a way of doing things (or specific gene expressions) rather than genetic specificity. A strict socialization of the members of a flock is necessary to maintain the reproductive praxis and therefore the species cohesion. The minimal flock is not a breeding pair but is specific to the type of reproductive arrangement. The secret of proper apportioning of gene (how many allele numbers should be there *etc.*,) is governed by mathematics, and this mathematics is specific to the reproductive arrangement. The biparental co-adaptation mechanisms held up by Paterson must be enlarged to include this mathematics of gene flow. Central to species is not specific mate recognition but the recognition of specific ways of organizing reproduction (I see parallels in human society such as matrilineal, patrilineal, matrifocal *etc.*). Each reproductive arrangement has its inheritance laws, be it for apportioning song, mate, territory or genes. The specificity of species is due to specificity of reproductive arrangements.

By now it will be clear that I am proposing a view of Mendalian genetics that runs contrarily to what prevails in contemporary thinking on speciation, especially among biologists such as Dobzhansky and Mayr who deduced a speciation model based on the nineteenth century biologist Gregor Mendel's theory of genetics of inheritance. Mendel cleanly misunderstood the function and nature of intergenerational gene flow and his disciples such as Dobzhansky (Dobzhansky 1951 pp 3-18) reduced speciation to a Mendelian population genetics. My strongest objection to Dobzhansky is that he assigns an apriori existence to a Mendelian population, while in truth such a population is an outcome of a speciation event (pace speciation not prior) and is an attempt to counter the loss of genetic specificity. Each species must have its specific Mendelian mathematics. In such a Mendelian population the law of inheritance takes the place of the inheritance itself so that species diferentiation survives and the genetic structures do not collapse into undifferntiation (into the hole in the middle made by the nulled gene complex).

In a critical or Mendelian population all the members have the same genetic combination (albeit reshuffled in the form of alleles) but the stock is maximized (optimized) only in one member by an optimum route of inheriting the stock (perhaps comparable to having the strongest copy). But the catch is the member in whom the stock is optimized is the weakest link in the route of transfer because (s)he is bound to dilute the stock after him (see appendix 4). The optimized member is rather inrrelevant to the survival of the stock. In such a case a better option is to go to one or both weaker (less optimized) copies (either at an earlier level of the family tree (a previous generation) – as do Tickell's flycatcher, magpie robin, babbler - or later level such as perhaps second copies coming from the same parents - iora; some species such as fish may be adapted for incest, snakes for homosexuality, but sadly this is not possible among bipedal humans) that still can maximize the stock in the next generation. It appears that such side stepping is required at regular intervals depending on the number of alleles or stock genes.

However I must here add that there is more to the Mendelian law of inheritance just as yet. It must be pointed out that actually there are two routes of optimizing the stock, what we may term as a legitimate route and an illegitimate detour. The former is got by mating between members who have two separate lineages (or family lines that means minimum in-breeding) but of the same stock as for example in ring species or clinal populations; the latter by mating between members that come from within a lineage (same family line that means maximum in-breeding) as for example in clannish populations (see Jinks, 1983; Mitton & Grant 1984 for a similar proposition labeled as heterozygosity and homozygosity but the latter is thought to be got by out breeeidng or dispersal). In the first case there will always be members in the stock who haven't contributed at all to any one of the lines, in the second there is no member (or very few members) in the critical population who has not contributed to the line. Both yield equally optimized copies but the comparison between them being that of a photo - legitimate - to a negative film – illegitimate. There are great many differences between the two copies even if genetically they are similarly optimized in the sense that this optimized member shares at least one allele with every member of the population. The hitch is that the legitimate member is genetically so conservative that even though he is geared to reproduce he produces a great quantity of weak copies (*i.e.,* a line of off springs that are genetically least optimized) that continue to remain so after him for a long time and definitely none of them or he produce any or little song (or such a novel adaptation). The illegitimate member can produce only song; if there is an offspring it remains reproductively inviable.

The solution of course is the same either way, side stepping up or down the line combined with differential infantile feeding. Because effectively speaking it is the penultimate member in a line who is the best optimized copy if not the most optimized. Most of the time, in actual practice species populations exhibits both the types of intergenerational mating sequences, *i.e.,* legitimate and illegitimate. Not only that a legitmate line can become the illegitimate at a later stage and vice versa. In fact – if we consider that populations practice cyclic and embedded mating sequences of the above sorts – members at the beginning of a illegitimate line will resemble that of the legitimate line because of fewer numbers of mating at that beginning level in the sequence. All this is directed to optimizing the mating chances of each member in the critical population. Even though down a legitimate line assortative mating may be followed strictly, members especially down the illegitimate line do not exercise mate choice but merely mating chance. Mating chance may be limited only by endogamy and incest, the latter applying in full measure only to humans. Even though it may be added that down the line in a Mendelian population secured by a fixed number of members the choice of mates by logic is greatly reduced and can affectively start to resemble assortative mating. Thus practice of such two different mating sequences in the same population explains why there is a singing male and a breeding male in mutually exclusive or distributed roles in a reproductive flock. Our song bird must be such a member that's why it is not important for him to reproduce. That's why songs turn out to be negative or signals of a lack and are kin (group) selected signals. It is the fruition of the species stock but the person inheriting it is the weakest link in its perpetuation.

If the population has not reached its critical level (or when a population has lost a breeding member or given it up to another sister stock) then it may be possible to import a new strain from a sister stock (subspecies). In fact there are species that do this at regular intervals (leaf birds, tree pies) even though this will not lead to a new subspecies it can in time relocate the population (change in clan head for example). The optimum route of inheritance for a species will be in the shape of one arm of the hour glass. At the apex such divergence of stock can occur if the stock chooses either of the optimized members as the scion. The best way is to retrace on the family line a bit and reproduce. Because effectively speaking it is the penultimate member in a line who is the best optimized copy if not the most optimized.

In a Mendelian population the importance of a member is framed by its role in maximizing the stock and a weak member is more important than the strong or best copy, especially in reproductive advantages. The evolutionary process is based on this logic rather than the law of survival of the fittest. Infanticide among many species, and that which became central to the group selection controversy in the '50's, I suggest is practiced on the optimized member (breeding or displaying both) by the group. Among birds eggs are always lost or hatchlings die. This must be a way of stemming the dilution of stock or its relocation should the member survive. But this appears to be possible only in younger species because as a species matures in time the optimized member cannot be killed like this. It survives by being genetically optimized and can cause major reorganizion in the family structure. Lastly, it may be noted that among species that are not known to have elaborate song (and other displays) could be merely that this optimized member is being eliminated like this and therefore can always develop song at some later stage. This may be true for all neoavis birds and not merely passerines.

The above observation therefore holds for both passerines and also non passerine mating systems. In fact I suggest that it has explanatory potential for understanding the cuckoo mating system which involves brood parasiting. The cuckoo being a strongly monogamous species tends to maintain only the legitimate lines in distinctive populations. Consequently both the breeding male and female can become totally conservative. The species thus may not produce any song or breed viably after a few generations even after taking recourse to side stepping along the line. So the cuckoos undertake to parasite on other species bird's far smaller than they. But interestingly it may be observed that the hosts for a male and for a female member are of different kinds. The male is hosted by species such as little flower pecker, the female by an unlikely host such as changeable hawk eagle and forest eagle owl. Even though in both situations the hatchlings usually suffer from early nutritional stress, in the male it leads to the development of species song capacity but in the female produces plumage morphology that mimics the host because this occurs in the presence of the singing male who suppresses song. The cuckoo sexual dimorphism as well as polymorphism (where female cuckoos in the same population display a range of intermediary coloured plumage, a range of shades from brown to deep brown to dark black spots) may be explained by the practice of such differential hosting.

Empirical observations point to the possibility that the optimum route for each species that belong to a same family clade may involve distinctive intergenerational mating sequences even though not the out come, *i.e.,* the existence of a singing and a breeding male in the flock. So that it appears that they owe their specific identity to simply the different mating sequences (reproductive systems) they adopt that regulate the break up of a shared genetic structure in different ways, making way for adoption of diversity of gene expressions (compare Tickell's, magpie robin family trees with Asian Brown which resembles that of iora given in Figures 1, 2 & 3 above). This results in strong reproductive barriers and exclusive songs. Songs and other displays must be a critical mechanism by which such reproductive arrangements are initiated, organized and held together, *i.e.,* songs bring about species cohesion. The diversity in song types and their species specificity must be mechanism of organizing specific reproductive arrangement. A word about rivalry and contest among the species, my field study points to the fact interspecific relation between the five sympatric members from the flycatcher family Tickell's blue, Asian brown, Asian paradise, veriditer, and magpie robin is governed by the need to maintain species identity and seems to be a biological protocol. A clue to this is the persistence of the interspecies relations across geographical barriers. The differential feeding habits also seem to stem from the same species identity protocol rather than a resolution of contests. Even though this has a spin off, for the eco niche resources get distributed amicably. It is true that in such a sympatric/ allopatric community individual member species can exert influences on each other's way of doing things. But my suggestion is that the structures of dominance and hierarchy that interspecies influences may give rise to in a community of birds are merely apparent and do not have deeply genetic causes. Song matching of birds therefore can be viewed as index to the genetic inter and intra specific relation – kinship structures - and can not be found among totally unrelated members (see Table Fourteen below). This brings us back to the actual song mechanism.

Picture 11: A female white browed bulbul produces display in front of her natal family members and not her betrothed. She has to persuade them to let her mate with him

In the following I try to understand the actual socio-psycho-biological mechanism that underlies the production and reception of song, *i.e.,* how does the song mechanism work? How can it explain as to what happens to the song capacity of the other members, if songs are indeed of a species nature and potentially each bird in a flock has equal genetic capacity for song? I suggest that the mechanism of song works by suppressing the songs of other members and directs them to make a choice for reproduction. Closer the genetic relation between the lead singer and the member greater is the song suppression. In fact the complete or partial suppression of song of some members gives rise to choice between reproduction or singing. Each member participates in a reproductive praxis that fetches him or her respectable place in the flock.

My emperical field observation on intergenerational comtinuity of song among especially birds such as iora and magpie robin has posed the difficulty in understanding why the species typical song is not produced invariantly in every generation (and in every member) if it is truly genetic in nature. It has further brought to the fore the possible role of the lead singer as social tutor in transmitting the species typical song down new generations. The hypothesis on song suppression is supported by the inverse correlation we find between the transmission of song and the seasonal capacity of the lead singer in the flock. The species song is not transmitted in the generation when the lead singer does produce intense song and vice versa. This suggests that presence of song must be negative signaling and gives rise to its suppression among newly fledged members, especially when combined with reproductive inviability of the lead singer social tutor. The social tutor must play a symbolic role in transmitting the song intergenerationally rather than actually "teach" the song.

Even though brain localization studies have investigated a great deal on the song production and song recognition mechanism of birds (Zeigler & Marler, 2012) there is still a long way to go. Such studies still assume that the bird sensory perceptions are organized similarly to that of humans and therefore there is insuffecient understanding of how a bird perceives the display signals. I suggest that the call for breeding song of a lead singer results in the suppression of the song of his mate and any other possible rival (for song) in his environment. He in fact suppresses the song of his female to such an extent that she can only opt to switch from song to another form of display (a kind of libidinous transfer): this explains why among most passerines for example iora, leafbird, bulbul, and oriole the female bird can be found rolling, swinging from branches and exhibiting such visual bodily display at the start of the breeding season but does not produce any vocalizations. (It is regrettable that in current ornithological literature these displays are attributed to breeding males. But this is far from true and moreover many of these birds are not sexually differentiable at this time of their season).

It is my prediction that owing to the small brain size of a bird, the way in which it is structured (for instance the location of the cochleal fluid) and its mechanism of sensory perception (for instance due to their small size animals see near object by longer focal length and vice versa), a repeated sound or a visual pattern has an effect of creating an illusion that it is the one producing it. Persistence of

Picture 12: Tippled sunbird

vision, and of other sensory experiences, has this effect of breaking the perception of distance between the object and subject in "bird brained" small sized creatures. This identification of the hearer with the singer produces a para somatic or trance like state in the brain of the hearer (perhaps caused by hormonal changes – reduction of corticosterone and testesterone for instance - induced by the listening). This leads to an immediate and momentary suppression of song sufficiently to give the lead singer an edge as far as song is concerned. My hunch is that all kinds of display of birds must be understood in this light.

This can also explain satisfactorily why fellow members in a flock produce different songs and show variations in song capacity. The song suppression mechanism functions in a way that lesser is the genetic distance greater is the suppression and greater is the variation in song capacity. Let me work out how and what I mean by lesser the genetic distance between the bird that sings and the one that hears greater is the affectivity of the mechanism of song suppression. In the instance of less genetic distance between the two members the song recognition is maximal so as to create the greatest break down of subject object differentiation. This leads to optimum song suppression response. But still song learning and production of a different song is also aspect of song suppression mechanism that differs only in degree and not kind. Among genetically divergent species that do not share even the niche, thus we do not find any or minimal song interphases (see discussion and Table fourteen).

As for my claim that song suppression works to direct the flock members to make a choice for reproducing: The mechanism of song suppression or song resolution mechanism must come into existence because the song of a bird acts as a negative signaling for the younger (usually the case but not always) breeding pair. The younger bird recognizes his song in the singer's and psychologically resists it. This makes him to turn to reproduction away from song. That is why most birds (especially breeding members) produce songs not mating initial but after the hatching. During the mating and the post mating stages the song mechanism is not available to the breeding members by the negative signaling-song suppression mechanism. It is in the post hatching when the fledglings are out, a maximum amount of vocalizations are produced. This also suggests that songs are addressed to the new chicks and not a communication between the mates themselves–I would even suggest that songs are the avian version of cradle songs and can be differently addressed to the male and female fledglings. This explains the sex differentiated production of song among members. The negative signaling song suppression hypothesis can also similarly throw light on the inter species dynamics of song in a local community of birds. The many kinds of the song matching found in a local population must be an outcome of the same mechanism; For instance the reverse song type matching among sympatric species and the elaborate song type matching between allopatric members. Most importantly it explains the evolution of song learning (usually by the lead singer and especially of other species) among birds.

From the point of molecular insight it is possible that song mechanism is located on a deleted/depleted gene by a novel 'pleiotropic co-adaptation' (co-adaptation meaning epistasis) of a gene meant to regulate reproduction on the new organism. Thus song is response to the same biological protocol as breeding. But since it is located on a deleted gene it functions as negative signaling as do elaborate plumage and other morphological appendages. In the lead singer there must be a relocation of the reproductive mechanism on the deleted gene and the song on the reproductive gene by the same 'pleiotropic co-adaptation' of a reproductive gene and a deleted gene. Even when the breeding and singing roles are not distributed among different birds the same mechanism must work even though at some stage in the reproductive process. Such a versatile novel mechanism can also explain song learning, especially why a species specific song can be erased and replaced by sounds from the habitat. This same mechanism could also explain loss of 'specific' morphology such as tail streamers (drongos, flycatcher, parrot) and molting (iora) during breeding. I suggest even sex difference among birds.

I have said that the singing bird makes his choice of song at the cost of his reproductive privilege. There is collateral to his sacrifice in the song suppression of his fellow members. The complimentary genetic expression of song and reproduction in a species flock must be triggered by the lead singer in the environment. Song of the lead singer redirects the libido of the other members towards gainful reproductive participation. He imposes asceticism on them to secure reproductive success for the flock. This is the dynamic of song within a species population. It is underwritten by a reproductive economy which is the reflection of nature's asceticism. This altruistic dynamic of song explains the genetic continuity of song capacity among a species.

I have presented a framework based on ultimate causes that explains the empirical observations I gleaned from my field study. Below in Table fourteen I have attempted to present the possible correlations for the various kinds of song matching (or song interphase) that I observed in the Bhadra region. My data suggests that there are basically two kinds of song interphases among members in a local population: production of dialects and production of geographical varieties. The former is found between genetically related members (intra species or subspecies or same family clade) and seems to be correlated to the genetic make up of the member. In many instances such songs seem to reflect even fine genetic differences between allele members in a same reproductive flock so that members in a breeding coalition display different but related songs. Such song matching involves structural differences (what may be termed as systemic shifts) in the song such as patterning of the notes/syllables (break up or reorganization of note order, introduction of new notes) and tonal variations; *e.g.,* reverse song type, repertoire song type-2, sex differentiated song, production of a more elaborate song *etc.* The latter kind of interphase, *i.e.,* geographical variety, reflects the allopatric (habitat) configuration and is mostly pure imitations that do not involve structural changes. Such geographical variety song interphase may be found even between genetically unrelated members, *e.g.,* among tailor bird and robin, rose ringed parrot and jungle babbler, leaf bird and shikra/minivet, even among animals that do not belong to the same class leaf bird and squirrel, or produced by gadgets such as radio or car horns, so on. The former type of interphase appears to be a dimension of the identity function of bird song while the latter territorial claim function, so that usually birds produce both type interphases in the same song delivery. Strictly the latter, *i.e.,* the territorial claim function-geographical interphase is produced by members who have completed at least one successful breeding season. Younger members in their first season do not produce such imitations. Such imitative content may serve as a description (mapping) of the territory and often produced in the form of a narrative sequence (story) (See spectrogram of the magpie robin peak breeding second season song in Figures 3a& b, Chapter two).

Even though it may be safely generalized that birds that are genetically related are more likely to engage in song matching several puzzling questions confront us. Why is there song type matching between only sub species pairs of leaf

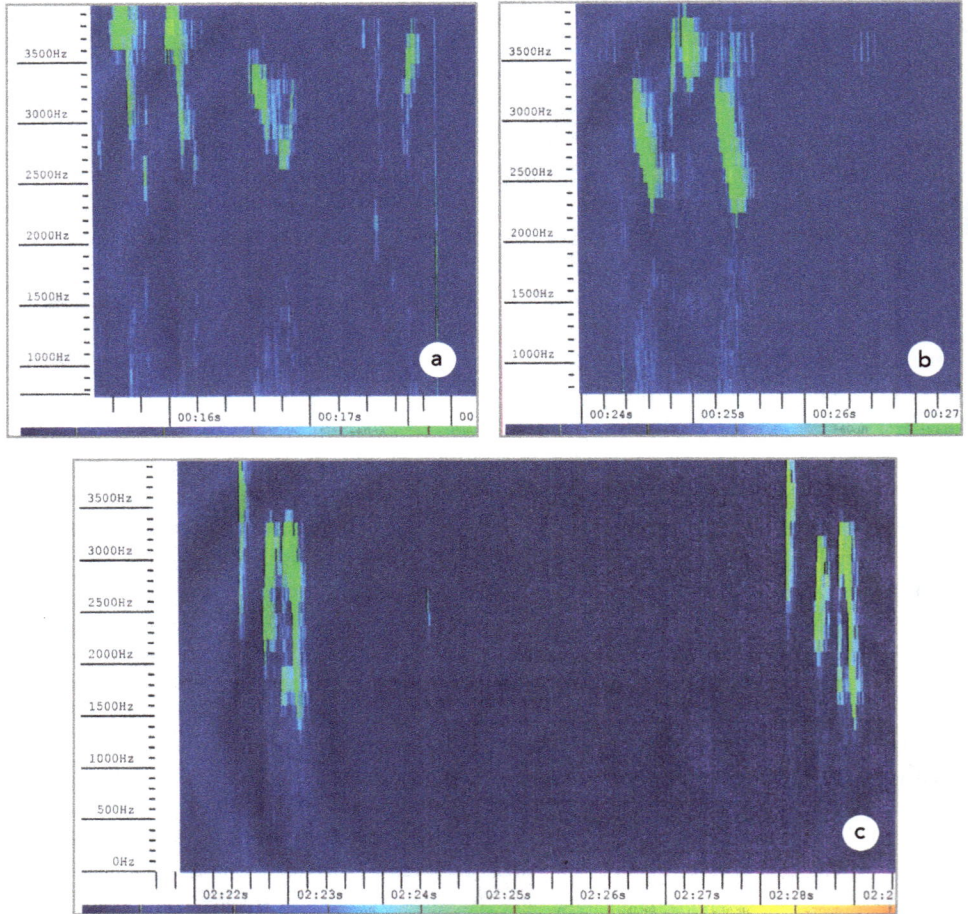

Figures 4a, b & c: Comparsion of imitations of the notes of the bulbul typical song. The king drongo (tweeter) and magpie robin (tweeter & twee tweeter) both produce notes comaparable to the bul bul (non breeding) (tweet tweeterio), respectively

birds and why do these reverse song type matching be so different among them? Why is there repertoire matching among treepie subspecies? Why should there be no song type matching between magpie robin and Tickell's blue flycatcher as both are classified in the same family muscicapedea? Why is there no song matching of any kind among the flycatchers but reverse song type matching among two species of sunbirds? Why does the red whiskered bulbul produce repertoire singing and not the other bulbuls? Where does the tailor bird fit in? Why is the song of the scimitar babbler so different from members of his family? Why it is among the non passerines such as barbets and doves there should be similarity in song type but no song matching is observed? My hunch is that these apparent anomalies suggest that song type matching does not merely reflect the genetic distance but the type of speciation relation and the hierarchy of genetic divergence: That is to say allopatry or ecological understanding of speciation is insufficient to explain the genetics of species. And that allopatry is merely a relational phenomenon that throws light on the nature of species existence/relation rather than speciation. And most significantly songs are fundamentally social in character. Song matching reflects the social organization of kinship structures.

Thus in the table below (and in the book) the ecological labels have not been used as speciation terms or even speciation processes (Chandler & Gromko, 1989) but as species relation or type of contact in most cases cited here secondary contact zones. There can be observed different kinds of contacts varying in degrees of intensity of intergrading. Among repertoirie singer type-2 we find maximum intergrading that involves interbreeding and such species are therefore conspecifics. Population differences (morphology and song type) arises due to differential allopatric nesting that uses very very local allopatric barriers such as food, type of nesting, vegetation and ground type, altitude, temperature, light and humidity and configuration of organisms. Sex differentiating complimentary breeding, as also its intergenerational differentiation give rise to reproductive inter and intra specific barriers among conspecific members - assortative mating- of such populations, *viz.,* among Tickell's flycatcher, magpie robin, tree pie, iora, sunbird so on. It also produces difference in song capacity and un typical song type, mainly repertoire singing, in such members only. Reverse song type matching on the other hand seems to correlate to different responses to similar habitat (with micro differences) but within the same range; but with opposing gradient of expansion, *e.g.,* leaf birds, sunbirds *etc.* This is another type of intergrading. Song type matching 1 and 2 do not fall within this category because it is among members in the same territory (type 1) or purely allopatric relation (type 2). Perhaps we can predict the spectrum of genetic distance and the correlation of song matching as shown in Table Fourteen given below. Outside this spectrum no song matching phenomenon is observed examples barbets, doves, Shikra and other birds.

Since the 1950's a lot of research based on field work has been undertaken in order to understand the emergence of differences and similaries between songs of birds and its relation to (or possible role in) the genetics of speciation under the rubric of dialect studies (used interchangeably with geographical variation). Initial field work (for example on the white crowned sparrow, Marler & Tamura 1962; Baker, 1975) seemed to have favoured the genetic explanation but subsequent molecular research has not unambiguously corroborated the field observations (Marler, 1952; Handford & Nottebohm, 1976; Yoktan *et al.,* 2011, *etc.*), leading to the conclusion that dialect variations may not always arise due to genetic differences. As a whole this has caused the emergence of a framework for understanding bird song in which genetics is given minimal explanatory role. Some half a dozen theories have been proposed which are labeled as historical model (Lemon, 1975; Slater, 1982, *etc.*), cultural model (Payne, 1978; Slater & Ince, 1979; Kroodsma *et al.,* 1999c) and habitat model (Nottebohm, 1975) *etc.*

Of particular interest is the discovery made by Nottebohm (1975) that variations in song can arise among genetically related populations due to acoustic adaptations to the habitat. Birds can acquire different song patterns with suitable acoustic qualities to the habitat type and such songs can appear as dialect

Table 14: Species Relation among bird species and its possible correlation with song matching displayed by them

Sl No	Type of species relation or type of contact	Type of song matching	Between Bird species
1.	Conspecific	Song suppression-1	1) Between male and female ioras of same clutch 2)Between older Tickell's & older members of other flycatchers
2.	Allopatry	Song type matching-1	1)Red whiskered bulbuls of different locations in territory 2) Two different generations of red whiskered bulbul
3.	Sympatry	Reverse song type matching-1	Cochinchinesis & frontalis ; NSR & insularis leafbirds
4.	Peripatry	Song type matching -2	i)Red whiskered, red vented & white browed bulbuls ii)Cochinchinensis & NSR
5.	Parapatry	Reverse song type matching-2	i)Between males of two kinds of sunbirds ii) Parvula tree pie female & pallida female
6.	Parapatry among conspecific (intermediary adaptations in conspecific)	Song suppression-2 (song mismatching)	1)Between male ioras of two generations 2) Older & younger tailor bird singing females
7.	Peripatry among conspecific (intermediary adaptations in conspecific)	Repertoire singing-2	1)Surrogating Magpie robin 2) Parvula Treepie male 3)sunbirds male
8.	convergent adaptation between nonconspecific	Repertoire singing-1	1)Tailor bird 2)red whiskered bulbuls (both imitate other species members in a territory
9.	divergent adaptation	Similar song types but no song matching observed	1)Between barbets 2) Between doves, 3)Between hooded oriole and treepie

difference. Nottebohm studied the trills patterns of the chingolo (*Zonotrichia capensis*) in Argentina and observed that a forest population produced slow trills, grass land population fast trills and those in arid habitat very slow trills. Perhaps some of my data can be explained by this last model, for instance the dialect differences between the two non intergrading leaf bird species golden fronted (cochinensis) and NSR (insularis), that occupy hill top and plain, respectively,

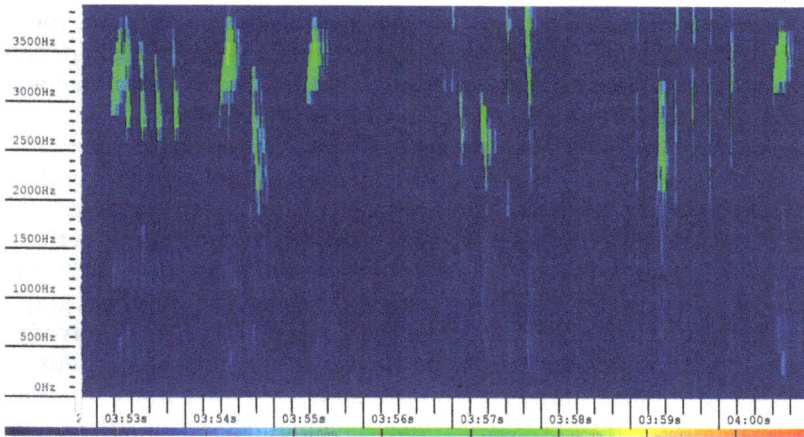

Figure 2: Spectrogram of the trill portion from the song of the leaf bird made up of imitations of other birds

chiefly regarding the whistle (not trill) portion of their song. In fact among leaf birds the trill portion is made of imitations of other species in the habitat but delivered as in a trill. Nottebohm's model does not provide a clue on the acquisition of such alterations to songs. Similarly reviewing my data on the iora in the same perspective, it seems possible that the song type variations in the lead singer and between the breeding male may be due to seasonal climatic changes in the environment such as degrees of moisture in the air, etc. which deters the song capacity. But it suggests that such climatic changes are accommodated to maintaining a fixed structure of intra–species song dynamics by promoting differential capacities for songs resemblance the use of allopatric feeding and nesting practice. In the same light. On the other hand it may not explain why sunbird males develop two different song types in the course of their season. Like wise the differences between the song patterns of the plains (parvula) and the hill slope (pallida) tree pies, because both of these usually produce vocalizations amidst trees in thick forest clumps (on the slopes as well as plains) rather than from the top of trees like the leaf birds do. Similarly, it may not explain the song patterns among the three species of bulbuls, because like the tree pies they tend to cluster under canopy of forest trees. It is possible that birds that adapt acoustically still do it on the basis of inner protocols or at least must communicate different contents.

Any such dialect variation still seems to point to the nature of avian vocalization as being generative; suggesting an underlying rule based grammar. The above mentioned explanatory models I suggest merely address the issues of how to understand the external processes of change that can ocur in the specific language rather than providing a grammar of species language and explain the intrinsic (linguistic or communicative) rules for assimilating change by way of dialect. It may be pointed out that dialects imply systemic quality of vocalization, but usually marked by not major grammatical shifts but shift in minor or peripheral rules (from the source language), For one, what is viewed as dialectical differences could be differences in referential content of the vocalizations relating to the rules of performance and body language (intonation and stress mainly), as for instance information regarding the stages of breeding - egg may be merely refered to by shortening the ending of the sound pattern refering fledged member); doubtful sex differences in early devlopment stage may be vocalized by tonal ending, internal states such as molting, may be expressed by tonal variation on the sound indicating species identity, *etc.* In this regard, it must be added it is a pity that much of the research and all the proposed models are based on quantitative studies of the song data where variations in songs are translated into visual graphs rather than relying on the ear. In my data presented above I have sought to differentiate between two kinds of song interphases purely dialectical and the other geographical or allopatric variations (see above). Much of the variations that are observed in visual sonograms (even minor ones) could be merely geographical variations, *i.e.,* due to the presence of other sounds (even data cables or power lines) specific to that habitat and therefore must be considered as incidental (not rule based, remove the sounds song pattern changes) or merely referential content by tonal pattern. But on the other hand there may be influences which are based on rules of species specific grammar, the real dialects.

Much of the research and conclusions on genetics of bird song I summarized from the current literature in the above paragraphs has focused on what is called as study of dialect variation especially in contiguous populations, among conspecific and sub species populations. Dialect study is of a totally different nature from the study of song types I presented above (also in Table Fourteen) based on my own field work in this book. In dialect study the minimal vocalizing units of members with same song type in different populations are compared for micro differences. Such studies do not focus on the differences in song types or repertoires of a species. In fact there are no studies that attempt to understand any possible relations of song types to genetics and kinship structure among birds of population in the literature. By my field observations on song type variations as well as dialectical differences I would propose that songs and their varieties are derived from a deep grammar of avian species and therefore – not that there is a gene for a specific song but that given the nature of genetics as I have outlined in this chapter - must be viewed as the social correlation of the specific genetics as well as primarily expressing the genetic relations or kinship structure among the birds. The coherence function of avian song is an intergral aspect of its performance and the cultural may be merely the other side of the genetic kinship coin.

As for the research attempts to find experimental proof of the genetics of bird songs, they are of two kinds. One is sub species classifications based on the pair wise correlation between song patterns and genetic distance values between many kinds of passerines in two different regions separated by a geographical barrier (see Marten *et al.,* in Renner & Rappole (eds.), 2011 for an example). Genetic distance value is based on cytochrome b gene of the mtDNA, which are thought to be gene markers of phylogeny among birds. This seems to be in keeping with my insight into speciation as a negative phenomenon. It seems appropriate that species specificity should be located outside the nucleus DNA and perhaps is proof of my claim. Lesser the genetic distance value lesser is the match in song pattern expectation. Even though such studies are successful they are of no use outside species classification.

A second type of study is allometric, *i.e.,* the counting of alleles among members of a population or across. Most dialect studies have been done with this method (Baker, 1975 for example). But I suggest there is yet insufficient understanding of the relation between genotype and phenotype. For one allellic frequency is considered to elict song dialect differences but so far no laboratory result has testified it (Handford & Nottebohm, 1976; Yoktan *et al.*, 2011, *etc.*). My suggestion is that allele frequency should inversely correlate with differences in song patterns, because two allele members must be genetically closest and therefore will produce same song. Alleles are the product of intergenerational gene flow in a species population and goverened by the law of Mendelain population.

My study of songbirds has thrown immense light on the nature of species existence and subspeciation. It appears that critical population size is the integral aspect of species existence, and should a population grow beyond this critical size that species is bound to diversify in the form of subspecies so as to maximize the niche. Such population allele is the mode of existence of every extant species and it is a measure of the hospitable ecological conditions. My field data suggests that there are mainly two kinds of such diversity; viz clinal and that which may be recognizably distinctive subspecies populations. Whereever a species cannot be found in subspecies then they exhibit clinal diversity. Clinal populations can be continually interbreeding (as among magpie robins) or intermittently (as among treepies). Those assigned as subspecies are generally non interbreeding. But can be dependent periodically on the parent populations for the stock. Clinal populations exhibit repertoirie singing whereas unique populations exhibit different kinds of song type matching or dialects. But still inspite of all these distinction in actual field observation the distinction between clinal and subspecies appears to be merely allopatric or contact type rather than any deep cause. And therefore sometimes subspecies nomenclature can become unwieldy. There may be as many subspecies/ clinal population according to the diversity and hospitality of the eco-setting. In my book in sections on the various passerines especially I have taken up the issue of subspecies existence and scrutinized Ali's (Ali & Ripley, 2001) and other records of subspecies. Among ioras for instance I have suggested that the existing subspecies classification is wrong. It is possible that ioras exhibit clinal distribution along their range over a span of period rather than geographical. Similarly the Tickell's is identified as having a remarkably wide distribution but no subspecies by Ali and others. My observations indicate that Tickell's organize into a clan with several breeding members active in a season. It is possible that this distribution is clinal with a new clan coming into existence periodically (in a cycle of 3-4 years when the clan head is renewed) that expands along the species expansive range. Among the drongo I have challenged the existence of species such as ashy and bronzed drongos, suggesting instead that they are nothing but the racket tailed members in post breeding plumage. In addition I have proposed that the racket tailed exist clinally comparable to the treepies maximizing the diversity of habitat across their range basically due to altitude changes and types of trees. Lastly and most importantly there is evidence that speciation and subspeciation are usually cluster events. The member species distributed in a structured way (akin to microorganism colonies). In such structured distribution there is a socially organized heirachy or natural dominance structure.

It is my conclusion that the patterns of song matching found among a local community of birds gives us important insights into the nature of species existence in addition to genetic relation and speciation event. The inter-intra species dynamics seem to replay the drama of speciation relation that might have taken place when the species first diversified or originated (the conditions for new gene expression to have come into life I wrote of earlier). This drama seems to be genetically coded and eternally available to be retrieved and reenacted among birds of every local population. This I think is a clue to the genetic diversity in a local population, each local microcosm reduplicates the phylogeny. Songs of birds reflect this drama of kinship. The microcosm serves as crystal ball for the interested student to get at the real event.

This book was intended to be about the song birds of Southern India. But the actual focus of the book has been a small territory of about 230 to 400 sq km sq kms on the banks of the river Bhadra. Perhaps one would think the claim is a tall order and that the book fails to cover the diversity of song birds to be found in the entire South Indian region. I must agree that as my data emerged, since I have pursued a radical natural methodology for field observation, taking note of what appeared before me, there seem to be a bias towards the passerines in my data.

Picture 13: Purple Swamphen

Even among the passerines, it will be observed that birds more tolerant to human habitation are more available to such a kind of study. For instance the Indian Pitta is a passerine song bird with an elaborate song routine during breeding season starting June. Even though it is found in the territory I investigated, the birds are shy of becoming conspicuous and perhaps perceive humans as a threat. They prefer deep jungles. Similarly is the case of the Malabar whistling thrush, one of the most beautiful singers in the region and a seasonally migrating bird.Other birds, smaller sized such as flower peckers, weaver birds and munias, larks and bushchats produce a large quantity of vocalizations and perhaps study of which would throw alternative perspective on the songs of passerines. Some of them are especially found in large sized flocks and make up a good portion of the local community of birds unlike the small sized flocks of those birds I have studied herein.

Perhaps it may be suggested that the generalizations that I have made in this book cannot be extended to all bird orders. It is also possible that even among passerines it relates only to old world birds and a different framework may be required for species falling outside. Darwin bases his theory of sexual selection mainly on the accounts of songs and other displays he heard of large-sized non passerines such as moorhens/swamphens and ducks. His graphic account and conclusion (see The Descent of Man chapter sexual selection in birds) has become deeply entrenched that ordinary mortals would not dare to challenge them. No authentic research has been ever conducted to find out the truth of the Shibboleth he established. There are a number of non passerines that breed in our own region such as the purple swamp hen, common moorhen, coot, water hen, jacanas, whistling and spot billed duck possibly can provide a rich source of data for the study of vocalizations of birds. From what I have observed of them I have not found anything to contradict my conclusions on the passerines (see appendix 2). The vocalization capacity of the Malabar pied hornbill appears to be remarkable too. There are other birds Malabar Trogon (Darwin refers to the trogopan and his colourful plumage) is another bird rarely sighted and prefers deep jungle. Similarly another clade of non passerines, raptors such as eagle owl, serpent eagles, hawk eagle and kites are native to this region. It is possible this intriguing family of non passerines birds will reveal some more perspectives on bird vocalizations.

My study of representative birds of southern India might not be the last word on the song of birds yet still I do think it is a worthwhile step in unraveling the secrets of a world different from ours. I am aware that many of the things I have been saying contradict the established institutionalized discourses on birds, genetics and paleobiology. But then I feel that located as I am on the margins (I do have a bachelors in science) of institutionalized science I have the previledge of staying away from the pressures to confirm. The problem with institutionalized discourses is that it cannot survive without a positivist affirmation of a phenomenon. My insights on speciation and evolutionary process present a universe of beings in an eternal race against genetic pre-determinism, but a universe of beings that would rather embrace love, connectedness, creativity and joy. It seems nature has outraced man in her genetic race for now. Let us hope she will continue to outwit him and keep abreast of his technological chase.

If you recall at the beginning of this book I narrated the story of a king in ancient Babylonia. There I did not complete the story to its end and something of it is left out. It seems when the female slave the king's confidante and scribe finished writing a page she would show it to the king and he would then impress his seal on it. In the last days of his life King Solomon was troubled by a niggling doubt: at the back of his mind he had begun to suspect that the birds themselves had instigated him to give up their magical boon by throwing a carrot in his path, the gold mines. Or was it that they had merely put him to test? It happened that when the slave had finished writing the last pages of his dilemma she found the king already dead and his retainers were making preparations to bury him. There was a crowd around his body and she could not penetrate it however much she tried. So the last page remained without his seal. When one of the retainers finally saw her with the book he seized it from her and finding that every single page had the king's seal he thought he should preserve it. He failed to look at the last page probably because the ink was wet and it stuck. And the book survived for centuries even though historians doubted this story because it was without the king's seal. The history of the human race is punctuated with accounts to regain the link with nature that is ever threatened by our ambition and materialism. I see this book as another such attempt to establish this fascinating connection with the language of nature, even if of course there be no king's seal on its pages.

Appendix 1

Avifauna of the Bhadra region recorded from February 2011 to September 2014 in an area of approximately 40 km radius

Below in Table One I present data on the birds sighted from February 2011 to September 2014 in the Bhadra region. The area under observation is roughly of 40 km radius and has been divided into two main parts. These two parts may be distinguished in a fundamental way as one being predominantly rural agrarian and the other as more densely populated by humans, non agrarian and situated at the edge of deep forest. The latter type territory is of 230 sq km and makes the epicenter of the total area under observation. The former type encircles the latter in a skewed way and makes up the outer circumference. It is also low lying grassy open land in comparison to the latter which is mostly of raised hilly lay out covered with sparse jungle vegetation.

Table A1: Avifauna of the Bhadra region recorded from February 2011 to September 2014 in an area of approximately 40 Km radius

Common Name	Scientific name	Location	Residence Type	Nos.
Red Spurfowl	*Galloperdix spadecea*	EP	R	
Grey Francolin	*Francolinus pondicerianus*	C	R	
Red Jungle fowl	*Gallus gallus*	EP/C	R	
Grey Jungle fowl	*Gallus sonnerati*	EP	R	
Indian peafowl	*Pavo cristatus*	EP/C	R	
Lesser Whistling Duck	*Dendrocygna javanica*	C	B	
Spot Billed Duck	*Anas peccilorhyncha*	C	B	
Brown-capped-Pygmy woodpecker	*Dendrocopas nanus*	EP	B	
Common Flameback	*Denopium javanense*	EP	B	
Black-rumped Flameback	*Dinopium benghalense*	EP	B	
White-naped woodpecker	*Chrysocolaptes festivus*	EP	B	
White- cheeked barbet	*Megalaima veridis*	EP	R	
Coppersmith barbet	*Megalaima haemacephala*	EP	R	
Malabar Grey hornbill	*Ocyceros griscus*	EP	R	
Indian Grey Hornbill	*Ocyceros birostris*	EP	R	
Malabar pied hornbill	*Anthracoceros coronatus*	EP	R	
Common Hoopoe	*Upupas epos*	EP	R	
Indian Roller	*Coracias benghalensis*	C	R	
Dollar Bird	*Eurystomus orientalis*	C	R	
Common Kingfisher	*Alcedo athis*	EP/C	R	
Stork Billed Kingfisher	*Halcyon capensis*	C	W	
White-Throated Kingfisher	*Halcyon smyrnensis*	EP/C	R	
Black-capped Kingfisher	*Halcyon pileata*	C	W	
Pied Kingfisher	*Ceryle rudis*	C	B	
Green bee- eater	*Merops orientalis*	EP/C	R/LM	
Blue tailed bee- eater	*Merops philippinus*	EP	B	
Chestnut headed Bee- eater	*Merops leschenaultia*	EP/C	R/LM	
Common Hawk cuckoo	*Hierococyx varius*	EP	LM	
Indian Cuckoo	*Cuculus micropterus*	EP	R	
Eurasian Cuckoo	*Cuculus conorus*	EP	R	

Contd...

Table A1 Contd...

Common Name	Scientific name	Location	Residence Type	Nos.
Bay banded Cuckoo	*Cacomantis sonnerati*	EP	UN	
Grey Bellied Cuckoo	*Cacomantis passerines*	EP	W	
Drongo cuckoo	*Surniculus lugubris*	EP	R	
Asian Koel	*Eudymanys scolopacea*	EP	R	
Blue Faced Malakoha	*Phaenicophaeus viridrostris*	EP	R	
Greater Coucal	*Cntropus sinensis*	EP	R	
Lesser Coucal	*Centropus bengalensis*	EP/C	R	
Rose Ringed Parakeet	*Psittacula krameri*	EP/C	R	
Plum Headed Parakeet	*Psitticula cynanocephalus*	EP/C	R	
Spotted Owlet	*Athene brama*	EP/C	R	
Indian Great Eagle Owl	*Bubo bubo Bengalensis*	EP	R	
Forest Eagle Owl	*Bubo nipalensis nipalensis*	EP	R	
Dusky Horned Owl	*Bubo coramandus coramandus*	EP	R	
Indian Nightjar	*Caprimulgus asiaticus*	EP	R	
Rock Pigeon	*Columbia livia*	EP/C	R	
Laughing Dove	*Streptopelia senegalensis*	EP	R	
Spotted Dove	*Streptopelia Chinensis*	EP/C	R	
Emerald Dove	*Chalcophaps indica*	EP	R	
Yellow footed Green Pigeon	*Treron pheonicoptra*	EP/C	LM	
Pompadour green pegion	*Treron pompadora*	C	R	
Lesser Florican	*Sypheotides indica*	C	R	
White breasted water hen	*Amaurornis pheonicurus*	EP/C	R	
Water cock	*Gallicrex cinerea*	C	R	
Purple Swamphen	*Porphyrio porphyrio*	C	R	
Common Moorhen	*Gallinula chloropus*	C	R	
Common Coot	*Fullica atra*	C	R	
Common Greenshank	*Tringa nebularia*	EP	UM	
Common Sandpiper	*Actitishypodeucos*	C	UN	
Wood Sandpiper	*Tringa glareola*	C	UN	
Pheasant-tailed Jacana	*Hydrphasianus chirurgus*	C	LM	
Bronze Winged Jacana	*Metopidius indicus*	C	LM	
Eurasian Thick Knee	*Burhinus odianemus*	C	UN	

Contd...

Table A1 Contd...

Common Name	Scientific name	Location	Residence Type	Nos.
Black Winged Stilts	*Himantopus himantopus*	C	UN	
Little Ringed Plover	*Charadrius dubius*	C	UN	
Yellow-Wattled Lapwing	*Vanellus malabaricus*	C	B	
Red -Wattled Lapwing	*Vanellus indicus*	EP	R	
Small Pranticole	*Glareola lacteal*	C	B	
River Tern	*Sterna aurantia*	C	B	
Lesser Noddy	*Anous tenuirostris*	C	B	
Black Shouldered Kite	*Elanus caeruleus*	EP/C	LM	
Brahminy Kite	*Haliastur Indus*	EP/C	R	
Pallas's Fish Eagle	*Haliaetus leucoryphus*	EP/C	R	
Short Toed Snake Eagle	*Circeatus gallicus*	EP/C	R	
Crested Serpent Eagle	*Splornis cheela*	EP/C	R	
Eurasian Marsh Harrier	*Circus aeruginosus*	C	LM	
Shikra	*Accipiter badius*	EP/C	R	
Besra	*Accipiter virgatus*	EP	R	
Eurasian Sparrow Hawk	*Accipiter nisus*	EP/C	R	
Changeable Hawk Eagle	*Spizaetus cirrhatus*	EP	R	
Common Kestrel	*Falco tinmunculus*	C	R	
Amur Falcon	*Falco amurensis*	C	R	
Little Cormorant	*Phalacrocorax niger*	C	R	
Indian Cormorant	*Phalacrocorax fuscicollis*	C	R	
Little Egret	*Egretta garzetta*	C	B	
Western Reef egret	*Egretta gularis*	C	B	
Great egret	*Casmerodius albus*	C	B	
Intermediate Egret	*Mesophoyx intermedia*	C	B	
Cattle Egret	*Bubulcus ibis*	EP/C	LM	
Purple Heron	*Ardea purpurea*	C	B	
Indian Pond Heron	*Ardeola grayii*	EP/C	LM	
Malayan Night Heron	*Gorsachius melanolophus*	C	B	
Cinnamon Bittern	*Ixobrychus cinnamoneas*	C	B	
Black Bittern	*Dupetor flavicolis*	C	B	
Black Headed Ibis	*Threskiornis melanocephalus*	CC	B	

Contd...

Table A1 *Contd...*

Common Name	Scientific name	Location	Residence Type	Nos.
Black Ibis	*Pseudibis papillosa*	C	B	
Painted Stork	*Mycteria leucocepha*	C	B	
Asian Openbill	*Anastomus oscitanus*	C	B	
Wolly Necked Stork	*Ciconia episcopus*	C	B	
White Stork	*Ciconia ciconia*	C	B	
Black necked Stork	*Ephippiorhynchus asiaticus*	C	B	
Indian Pitta	*Pitta brachyuran*	EP	B/M	
Asian fairy bluebird	*Irena piella*	EP	W	
Blue Winged Leaf Bird	*Chloropsis cochinchinesis*	EP	B	
GoldenFronted Leaf Birds	*Chloropsis aurifrons*	EP	R	
Rufous tailed shrike	*Lanius Isabellinus*	EP	R	
Long Tailed Shrike	*Lanius schach*	C	R	
Rufous tree pie	*Denrocitta vagabunda*	EP/C	R	
House Crow	*Corvus splendens*	EP	R	
Large Billed Crow	*Corvus macrorhynchos*	EP	R	
Eurasian Golden Oriole	*Oriolus oriolus*	EP	R	
Black naped Oriole	*Oriolus chinensis*	EP	R	
Black hooded Oriole	*Oriolus Xanthornus*	EP	R	
Ashy Wood swallow	*Artamus fuscus*	EP	B	
Black Headed Cuckoo Shrike	*Coracina melanptera*	EP	R	
Rosy Minivet	*Pericrocotus roseus*	EP	R	
Small Minivet	*Pericrocotus cinnamoneus*	EP	R	
Scarlet Minivet	*Pericrocotus flammeus*	EP	R	
Bar Winged Flycatcher shrike	*Hemipus picatus*	EP	R	
Black Drongo	*Dicrurus macrocercus*	EP/C	R	
Ashy Drongo	*Dicrurus leucophaeus*	EP	R	
White bellied Drongo	*Dicrurus caerulescens*	EP	R	
Bronzed Drongo	*Dicrurus aeneus*	EP	R	
Spangled Drongo	*Dicrurus hottentottus*	EP	R	
Racket Tailed Drongo	*Dicrurus paraadiseus*	EP	R	
AsianParadise Flycatcher	*Terpsiphone paradise*	EP	R	
CommonIora	*Aegithina tiphia*	EP	R	
Common Woodshrike	*Tephrodornis pondicerianus*	EP	R	

Contd...

Table A1 Contd...

Common Name	Scientific name	Location	Residence Type	Nos.
Blue capped Rock Thrush	*Montocola cinclorhynchus*	EP	W	
Blue Rock Thrush	*Monticola solitaries*	EP	B	
Malabar Whistling Thrush	*Myophonus horsfieldi*	EP	B	
Orange Headed Thrush	*Zoothera citrine*	EP	B	
Scaly Thrush	*Zoothera dauma*	EP	B	
Eurasian Black bird	*Turdus merula*	EP	B	
Asian Brown Flycatcher	*Muscicapa daurica*	EP	R	
Rusty Tailed Flycatcher	*Muscicapa ruficauda*	EP	R	
Veriditer Flycatcher	*Eumyias thalassina*	EP	R	
Nilgiri Flycatcher	*Eumyias albicaudata*	EP	R	
Tickell's Blue Flycatcher	*Cyornis tickellia*	EP	R	
Oriental Magpie Robin	*Copyschus saularis*	EP/C	R	
Brown Rock chat	*Cercomela fusca*	C	R	
Indian Robin	*Saxicoloides fulicata*	C	R	
Pied Bushchat	*Saxicola caprata*	C	R	
Isabelline Wheat eater	*Oenanthe isabellina*	C	R	
Chest nut tailed Starling	*Sturnus malabaricus*	EP	UN	
Brahminy Starling	*Sturnus pagodarium*	C	R	
Common myna	*Acridotheres tristis*	EP/C	R	
Jungle Myna	*Acridotheres fuscus*	EP/C	R	
Great tit	*Parus major*	EP	W	
Dusky craig Martin	*Hirundo concolor*	EP	B/W	
Red rumped Swallow	*Hirundo daurica*	C	B/W	
Red -whiskered Bulbul	*Pycnonotus jocosus*	EP/C	R	
Red-vented Bulbul	*Pycnonotus cafer*	EP/C	R	
White- browed Bulbul	*Pycnonotus luteolus*	EP	R	
Ziting Cisticola	*Cisticola juncidis*	EP	W	
Grey beasted Prinia	*Prinia hodgsonii*	EP	R	
Jungle Prinia	*Prinia sylvatica*	EP/C	R	
Ashy Prinia	*Prinia socialis*	EP	R	
Plain Prinia	*Prinia inornata*	EP	R	
Oriental White eye	*Zosterops palpebrosus*	EP	R	
Grasshopper Warbler	*Locustella naevia*	EP	W	

Contd...

Table A1 Contd...

Common Name	Scientific name	Location	Residence Type	Nos.
Paddyfield Warbler	*Acrocephalus agricolla*	EP	W	
Blyth's Reed Warbler	*Acrocephalus dumetrorum*	EP	W	
Clamouraous Red Warbler	*Acrocephalus stentoreus*	EP	W	
Thick billed Warbler	*Acrocephalus aedon*	EP	W	
Booted Warbler	*Hippolias caligata*	EP	W	
Common tailor Bird	*Orthotonus sutorius*	EP/C	R	
Common Chiffchaff	*Phylloscopus colybita*	EP	W	
Dusky Warbler	*Phylloscopus fuscatus*	EP	W	
Greenish Warbler	*Phylloscopus trochiloides*	EP	W	
Striated Grassbird	*Megalurus palustris*	EP	W	
Broad tailed Grassbird	*Schoenicola platyura*	EP	W	
Puff throated Babbler	*Pellorneum ruficeps*	EP	W	
Indian Scimitar Babbler	*Pomatorhinus horsfieldii*	EP	R	
Tawny-bellied Babbler	*Dumetia hyperythra*	EP	B/LM	
Dark –fronted Babbler	*Rhopocichla atriceps*	EP	B/LM	
Striped Tit Babbler	*Macronous gularis*	EP	UN	
Yellow-eyed Babbler	*Chrysomma sinense*	EP/C	R/LM	
Common Babbler	*Turdoides caudatus*	EP	R	
Rufous Babbler	*Turdoides subrufus*	EP/C	B/LM	
Jungle Babbler	*Turdoides striatus*	EP/C	R	
Yellow Billed Babbler	*Turdoides affinis*	EP/C	B/LM	
Brown Cheeked Fulvetta	*Alcippe poioicephala*	EP	W	
Indian Bushlark	*Mirafra erythroptera*	C	UN	
Rufous tailed lark	*Ammomanes phoenicurus*	C	UN	
Malabar Lark	*Galerida malabarica*	C	UN	
Sykes Lark	*Galerida deva*	C	UN	
Thick billed Flowerpecker	*Dicaeum agile*	EP	LM	
Pale-billed Flowerpecker	*Dicaeum erythrorynchos*	EP	R	
Plain Flowerpecker	*Dicaeum concolor*	EP	LM	
Purple rumped Sunbird	*Nectarinia zeylonica*	EP	R	
Crimson Backed Sunbird	*Nectarinia minima*	EP	R/LM	
Purple Sunbird	*Nectarinia asiatica*	EP	R/LM	
Loten's Sunbird	*Nectarinia lotenia*	EP	R/LM	

Contd...

Table A1 Contd...

Common Name	Scientific name	Location	Residence Type	Nos.
Crimson Sunbird	*Aethopyga siparaja*	EP	R/LM	
Forest Wagtail	*Dendronanthus indicus*	EP	W	
White-browed wagtail	*Motacilla-maderaspatensis*	EP/C	R	
Citrine Wagtail	*Motacilla citreola*	EP/C	R/LM	
Yellow Wagtail	*Motacilla flora*	C	R/LM	
Grey wagtail	*Motacilla cinerra*	EP/C	LM/R	
Paddyfield Pipit	*Anthus rufulus*	C	R/LM	
House Sparrow	*Passer domesticus*	C	R	
Chestnut Shouldered Petronia	*Petronia xanthocollis*	EP/C	W	
Baya Weaver	*Ploceus philippinus*	C	R/LM	
White rumped Munia	*Lonchura striata*	EP	R/LM	
Scaly breasted Munia	*Lonchura punctulata*	EP/C	R/LM	

Legend: EP=epicenter area; C= circumference area; R=resident; B=breeding; M=long distance migrating; LM=local migrating; W= non breeding winter visitor; UN=unknown residence type. The nomenclature for the above table followsGrimmet and Inskipp (2007).

Table A2: Field Statistics* of Passerine & Non PasserineSong Birds Under Observation for This Book

Sl No	Common name	Scientific name	Range	Subspecies for			Subfamily/ genus	Family
				India	South India	Bhadra/ western ghats		
1.	Orange headed ground thrush	Zoothera citrina	All along the hill range of India, from Himalayas, the Gangetic plains along W and E ghats.	4	Z.C. Cyanatus	Z.C.cyanatus	Turdinae / Zoothera	M U S C
2.	Oriental magpie robin	Copsychus saularis	All of India upto 1500m of Himalayan foothills	2 (+1 for Andaman Islands)	i)C. s. saularis ii)C. .s ceylonensis	i)C. s. saularis ii) C. .s ceylonensis	Turdinae/ Copsychus	I C A P
3.	White browed Scimitar Babbler	Pomatorhinus horsefieldi	Spread all along the hilly ranges of Himalayas, western and eastern ghats	4	i)p.h. travencorensis ii) p.h. madraspatensis iii) p.h. horsefieldi	i)p.h. travencorensis ii)p.h. horsefieldi	Timalinae/ pomatorhinus	I D A E
4.	Tickell's red breasted blue flycatcher	Muscicapa tickelliea	Resident breeder from foothills of Himalayas to Kanyakumari, even the desert regions of Rajasthan and Kutch	1 (+1 for Ceylon)	M.t. tickelliae		Muscicapinae/ muscicapa	
5.	Grey breasted prinia	Prinia hodgsoni	All over India from 1800m Himalayas, but absent in W Rajasthan	4	i)P.h.albogulars; ii) p.h. hodgsoni	ii) p.h. hodgsoni	Sylvinae/ prinia	
6.	Tailor Bird	Orthotomus sutorius	All over India	3+ 2 for Ceylon	i)O.s.guzuratus ii) O.s. patia	O.s. guzuratus	Sylvinea/ Orthotomus	

Contd...

Table A2 Contd...

Sl No	Common name	Scientific name	Range	Subspecies for — India	South India	Bhadra/western ghats	Subfamily/genus	Family
7.	Common iora	*Aegethina tiphia*	All over Inidam from 1800m Himalayas, but absent in deserts of NW Rajasthan and Kutch	5	i) A.t. Deignani ii)A.t. multicolor	i) A.t. Deignani ii)A.t. multicolor	Aegithina	I R E N
8.	Golden fronted leaf bird	*Chloropsis aurifrons*	Spread all along the hilly ranges of Himalayas, western and eastern ghats	3	i)C. a. frontalis iii)C. a. insularis	i)C. a. frontalis iii)C. a. insularis	Chloropsis	E D A E
9.	Blue winged (Gold mantled) leaf bird	*Chloropsis cochinchinensis*	All of southern India south of Punjab, Rajasthan and N W Gujarat	2	C.c.jerdoni	C.c.jerdoni		
10.	Racket tailed drongo	*Dicrurus paradiseus*	All of forested India, from south and east of S Gujarat to Kumaon hills	4	D.p paradiseus	D.p paradiseus	Dicrurus	Dicrur-idae
11.	Black hooded oriole	*Oriolus Xanthornus*	Resident subject to local movements. From Himalayan foothills(1200m to kanyakumari	1 (+1 for Ceylon +1 for Andaman Islands	O x. xanthornus	O x. xanthornus	oriolus	Orioli-dae
12.	Indian (rufous) tree pie	*Dendrocitta vagabunda*	From Himalayan foothills (1500m) to kanyakumari	5	i)D.v. pallida ii) D.v .parvula iii)D.v. verneyi	i)D.v. pallida ii) D.v .parvula	Dendrocitta	Corvi-dae

Contd...

Table A2 Contd...

Sl No	Common name	Scientific name	Range	Subspecies for			Subfamily/ genus	Family
				India	South India	Bhadra/western ghats		
13.	Red whiskered bulbul	*Pycnonotus jocosus*	All along the hilly areas of Garhwal, south of satpura hills and western ghats.	5 (+1 for Andaman Islands)	i)P.j. fuscicaudatus ii) P.j.emeria	P.j. fuscicaudatus	Pyconotus	PYCNONOTIDAE
14.	Red vented bulbul	*Pycnonotus cafer*	All over India from 1800m Himalayan range	6 (+1 for Ceylon)	i)P.c. cafer ii)P.c. wetmorei iii)P.c. humayuni	P.c..cafer		
15.	White browed bulbul	*Pycnonotus luteolus*	only southern peninsular India, downwards from C Gujarat to southern W Bengal	1 (+ 1 for ceylon)	P.l. luteolus	P.l. luteolus		
16.	Purple sunbird	*Nectarinia asiatica*	All over India	3	No data availble		Nectarininae/ nectarnini	NECTARINIDAE
17.	Purple rumped sunbird	*Nectarinia Zeylonica*	All over south India starting from Bombay a line south of Madhya Pradesh, S Bihar and Bengal	none recorded	None recorded			
18.	Asian Koel	*Eudynamys scolopacea*	All over India but uncommon in the drier regions	1	E.s. scolopacae	E.s. scolopacae	Eudynamys	

Contd...

Table A2 Contd...

Sl No	Common name	Scientific name	Range	Subspecies for India	South India	Bhadra/western ghats	Subfamily/genus	Family
19.	Indian cuckoo	*Cuculus micropterus*	All over Inida except NW desert regions	1	C. m.micropterus	C. m.micropterus	Micropterus	CUCULIDAE
20.	(Eur)Asian cuckoo	*Cuculus canorus*	All over India except NW desert	3	i)C.c.canorus ii) C. c. subtelephonus	i)C.c.canorus ii) C c. subtelephonus	Canorus	CUCULIDAE
21.	Blue faced Malakoha	*Rhopodytes-viridirostris*	All over India in semideced ious ever green forests	2	Rhopodytes viridirostris	Rhopodytes viridirostris	Rhopo dytes	
22.	White cheeked barbet white cheeked barbet	*Megalaima viridis*	W and E ghats of southern India and niligiri hills	0	m.viridis	m.viridis	Megal aima	CAPITONIDAE
23.	Coppersmith barbet	*Megalaima haemacephala*	All over Inida	2(+1 for Ceylon)	m. haemacephala ii)m.r.malabaricus	i)m. haemace-phala ii)m.r.malaba-ricus		
24.	Spotted dove	*Streptophelia chinensis*	All over India	1 (+1 +1for Ceylon, Burma & NEFA)	s.c. suratensis	s.c. suratensis	Strepto pelia	Columbidae

Legend:*The data presented in the above table is extracted from Ali &Ripley (2001), Ali (2002, 13th edition) and Grewal (2000) and is not record of my field observations on subspecies recorded in the sections on the birds in chapters 2 and 3 of the book above."

Appendix 2

Darwin's theories of sexual selection and natural selection

It is from Darwin we have the theory of sexual selection, where he posits that the female of a species has the prerogative to select the fittest male for breeding. He attributes to features such as song, colourful plumage and appendages such as collar, comb, tufts, spurs and tails the status of secondary sexual features and as aiding the female in sexual selection. He gives remarkably lurid accounts of how such appendages are used in courtship and in male contest over a female. According to his accounts among many birds such as purple swamphen, musk duck, jungle fowl, spurfowl, lapwings and some other birds that belong to a clade cluster such as fowl, the mating male is selected on the basis of a physical contest among the several desiring males for the same female. He derives his theory of sexual selection from these accounts.

First of all it may be noted that even though he gave very lurid graphic accounts of sexual contests none of them are first hand his own but reports he gleaned from several places. I suggest that had he proceeded to investigate the accounts he would have come to a different conclusion. Let me start with the jungle fowl, the wild originator of the domestic fowl whose eggs we consume as part of our daily diet all over the world. Even to this day every now and again the cocks are captured from the jungle and interbreed with a domesticated hen in order to replenish the stock otherwise they would become extinct. Now it is common knowledge that in rural countryside cock fights between the cocks of the jungle fowl is a favourite sport. In the sport two captive supposedly male fowls are set upon each other to fight. Invariably they fight until one of them is killed. Darwin refers to the cock fight as proof of the sexual selection theory and jumps to the conclusion that the cocks are fighting over a mate. As I have noted in chapter four of the book my first hand observation suggests that the bird with the crest is a female on her way to completing her breeding career over a fixed number of seasons. This course is accompanied by the developing display plumage and neutering. So the fight is actually between two neutered females and the fighting cock merely a folk mythology. Perhaps this fight between two neutered females is a duplication of a pre-mating contest between a breeding pair and surely not between two rival males. In addition it is not misplaced here to remind that among placental animals such as dogs, cats and hyenas the same female is impregnated by several males in the same breeding event so that there is no question of rivalry between the males nor is there mate choice on the part of the female.

Majority of the water fowls, such as purple swamphen, common moorhen, and common coot are nidifigous in breeding, meaning they do not lay eggs but straightaway produce younglings which have attained body hairs and can swim straightaway in the water from the minute of birth. The female among them holds the eggs in her womb till the last minute and lets them hatch inside her. This is because in their eco habitat they are threatened by predators like eels and it would be too much risk to incubate them and hatch them outside in a nest. Large heavily padded nests are built floating in the water for the female to sit in during the gestation period when she might feel weakened when the birthing is taking place n her womb. But the newly born chicks straight away enter the water and follow after the parents and start to feed. This is the best way of maximizing their life chances.

Many of such nidifigous birds are polyandrous with one female having several males. The males come handy in steering the newly born fully fledged younglings that are still very small in size and need care and child minding. Moreover there is usually a batch of three of them produced one after the other. The contest over mate that Darwin accounts in his Descent of Man (chapter XIII: Secondary Sexual Characteristics of Birds) I suggest was not a contest but a breeding male wooing/enlisting the help of another male as co mate in this serious business of reproduction. Generally once the chicks are out they swim in between three adults in a triangular formation. Very often chicks are transferred from one lake to another at a certain age I suggest to teach them of the alternative nesting living sites, food resources and familiarize with the territory but mainly to teach them to fly. While flying from one site to the other they fly again in a triangular formation. In my view Darwin theory of species and evolution suffers from anthropocentrism combined with a most dangerous brand of sexism. And Darwin is replete with masculinism for instance when he makes statements such as "male is usually larger, or is stronger or is older than the female." This is in fact not true in the least. Among animals, birds especially, the female is very often older, stronger and larger in size, *e.g.*, among owls and other raptors, among passerines such as iora.

Darwin considers beauty, such as bright plumage and appendages as an important attribute that sways the female's choice. I suggest that he quite wrongly understood the function of beauty in nature. Field observations lead one to conclude that the effect of beauty in nature is to stupefy by creating optical (and other sensory) illusions closer to the concept of aposematism (proposed by Alfred Russell Wallace, Darwin's colleague and Edward Bagnall Poultan after him; see Komarek,1998 & Ruxton *et al.*, 2004 for an account). In addition as I pointed out in chapter Four of the book, theories about displays are argued wrongly on the assumption that bird and other animals have visual and other sensory organs function similarly to that of humans. Not only that a recent study on bird feathers has revealed that not always the colour is due to pigmentation. For instance the peacock covert is made of tiny colourless globules that refract light like a glass prism. The colour display can be changed depending upon the angle of refracted light. So the effect of the tail covert display on another member would be unpredictable to a third viewer. It is a good hypothesis that animals produce signals suited to eliciting typical responses from the other member, that is akin to "conning" or fooling. By this among birds stunning display of plumage and song is used to redirect the libido of a young breeding pair to the more serious and responsible activity of reproduction, raising the young ones. It is an instrument of psychological control or intimidation deployed by an adult on a younger pair just entering into a sexual relation; negative signaling as I have called it in the book.

This is also true of smells and fragrances we find in the wild. The musk of ducks and other animals like deer are meant as herding devices to keep a reproductive flock together rather than for mate choice. Among deer the antler is not the breeding male. He is actually a neutered male around whom a flock is herded. The antler male loses his reproductive capacity when he grows his antlers. Among other males the antler development is arrested when they enter breeding. I believe the heavy antlers of the non breeding male is a symbolic control mechanism among deer warning them against egotism and rivalry which is sure to bring extinction. Among the hyenas a non breeding female's hymen hardens sufficiently so that she can project it outside and make it appear like a penis. She is said to have a pseudo penis. She in fact does not have any unique organ called as that, but it is only her hardened hymen. The pseudo penis develops because she is used by young pubescent males in her flock who become sexually aggressive in order to avoid mating with the actual breeding females before they reach reproductive maturity. Her displays of the pseudo penis before a young pubescent female just entering into a reproductive role serves as a reminder to the younger female of the hazards and turns her away from indulging sexually towards more responsible role of caring for the young. The same phenomenon may be noticed among the Rhesus (bonnet) monkeys in the Bhadra region. In a hoard there is always a female who can display her hymen in much the same way as the female hyena. In the plant kingdom fragrance that newly bloomed flowers emit repels the bee and other insects which are possible pollinators. It keeps them at bay until the flower is mature and ready to be pollinated/or yield pollen. Not only that when the fragrance evaporates it creates a vacuum around the flower to which the insects are drawn. The fragrance minimizes the chance of becoming fertilized/transfer pollen before the zygote is fully mature and it requires some time after the flower opens to the sun. A prematurely fertilized flower will not yield viable seeds and merely falls off the plant. The colorful petals of a flower are probably meant to give the insects camouflage while sucking the nectar and allow sitting sufficiently in the flower to transfer the pollen.

Many of the water bird species have red wattles on the spot between the beak and forehead. Some of them have yellow wattles. All of these birds with wattles are not distinguishable sexually by morphological features. Both sexes have the wattles. The colours red and yellow has the effect of producing blind spots in the viewer causing temporary blindness. Birds such as swamp hens use the wattles in this way to ward of predators. It is foolish that any one should think that the wattles are there to attract a female. The whistling duck breeds in the Bhadra region in winter. They are coterie breeders with several pairs putting their lot together for the sake of raising the young. Coterie breeder means young nidifigous chicks from several pairs are huddled together in a pen made up of floating vegetation with younger males seeking mates posted strategically around them. The ducks produce oil from a gland to rub on their feathers which I believe repels the water and makes them water proof. They are highly sociable birds. It is the privilege of the in- breeding pairs to share the ritual of rubbing the oil on their feathers and swim in the water while the younger males seeking to have a female from these breeding pairs do the business of babysitting. It is hard to believe from what I saw of the ducks that the musk duck that Darwin refers to should indulge in bloody battles over a female. In The descent of Man he recounts an episode where a Mr Bartlet reported a lot of feathers he found in the lake water which he believed came from such a battle. My hunch is Mr Bartlet found molting feathers and wrongly attributed to a sex battle. (It is possible that Mr Bartlet inadvertently caused a molting epidemic among the ducks and other water birds in the lake by attempting to feed them with salt water fish out of ignorance.) Most large sized birds when they molt start to shed feathers in clumps sometimes with bits of upper skin attached to them like scabs from wounds. This could be because molting produces itching in the bird and leads to lesions. In my field trips once I found a territory which nested owls and other raptors strewed with clumps of feathers. In no time I traced it to a young serpent eagle who came back to the site every evening.

As far as the proof Darwin gives for natural selection and survival of the fittest it is riddled with ignorance and misguided judgment about the habits of birds, other animals and the vegetation kingdom. For instance in The Origin of Species he cites an example where a certain species of firs have become extinct for being cropped by cattle. In my observation species that are food to other species always thrive better, for they reproduce or rejuvenate faster and produce more in order to compensate for the loss. Grasses and trees frequently cropped by grazing animals such as deer and sheep always grow faster and more profusely than those which do not make up the diet of some animal or other. Similarly birds such as doves which make up diet of many animals reproduce in larger numbers. If a plant or an animal is not thriving in a region it is more likely that they are either harmful to a species or useless and so do not grow. But there are few plants which do not find some use or the other. In fact older animals are known to crop down poisonous plants and vegetation during breeding time in order to safeguard the young from feeding on them. It is likely they produce different kinds of saliva while cropping which either retards or accelerates the re-growth of these herbs. Monkeys for instance strip the guava trees of fruit before they ripened when there are younger members in their herd to forage on the tree. The guava fruit is made up of small seeds which cause indigestion and constipation that leads to death of young monkeys. The ripened sweet fruit can make the young monkeys greedy and overeat. But on the other hand many of the animals can allow some poisonous or thorny plants to thrive to such a vast expanse of their regular territory so as to form a barricade against predators from reaching them too fast. Most poisonous plants are used as medicines against infections and may be allowed to grow here and there for the purpose along the animals' trail. If a particular species seems to thrive more than another it could be merely that that species is in the food chain of another in that niche or on the other side could be poison to the other, but and not the case of survival of the fittest or competition for life as Darwin terms in The origin. Food, reproduction and nesting are three necessities that everyone in the animal kingdom can understand and are accommodated without contest. There is a better chance of surviving by mutual co operation than contest over such fundamental needs. Survival ultimately is all about surviving together. A niche may be distributed by developing complimentary habits, *i.e.,* diversifying.

Adaptations are mark of creativity of organisms and are idiosyncratic, rather than for selection and fitness. Secondly, nothing in nature is static, and everything is in a flux, even the best adaptations. The speciation process is a time bound bio-mechanical clock ticking relentlessly away. No doubt it is the way nature rejuvenates itself continuously and it spells change. This relentless rule of flux ensures that even the strongest in the end will perforce be nudged out of its niche to make way for the next: perhaps life in nature is more comparable to a game of musical chairs or kho kho. The law of numbers or mathematics is the only law that ultimately governs life on this earth. What appeared to Darwin as selection may be only the mathematics underlying species existence: How many numbers of minimum alleles are required for the species to exist. or should be the critical size of the reproductive flock? This would determine why some species are more profuse and are in more numbers; Likewise, the absence of intermediaries among some species. This will no doubt create a structure of dominance and hierarchy but not due to selection or survival of fittest between them but by nature of their specific existence. Even the existence of hybrid forms in some species could be a feature of their species, serving as sentinels of a species uniqueness.

But then I think all life on this earth is held together in a web of relations and participate in a kinship structure, subject to nothing else but the laws of algebra, an insight had by ancient philosophers like Plato. That we are together and can never be alone is the only compensation.

Appendix 3

A note on methodology in ornithological research

Evolutionary science and genetics has come a long way since Darwin. Any one reading Darwin would undoubtedly find fault with the methodology he pursued in order to come to a conclusion. None of his empirical accounts are first hand especially in The Origin and The Decent (and my accounts are all first hand eye witness). He accepts accounts of the phenomenon as well as its explanation as given to him by men most of them involved in cattle rearing, or colonialists in countries like India and Africa who may be merely exoticizing and being racist.

During his time already there were many stringent laws barring against felling of trees, hunting of wild animals in Britain as well as the colonies, *etc.* (In India Forest Acts became established by 1832). The barons and aristocrats were the only people allowed to hunt in the wild. The laws perhaps did not serve purely the purpose of conservation because the royalty was finding it more and more difficult to find good game and this brought about strict legislations. A rustic who had as much taste for deer meat or a boar, had to justify the killing to agents of the law. Moreover in the colonies there was a flourishing trade in exotic birds and other animals stuffed of course, musk duck including that Darwin devotes so much attention to. The catching interest in evolutionary biology must have added to the illegal hunting. Killing license was given only when wild animals became a threat to human life. So I suggest that it became important, if a bird or any animal was killed, to establish to the authority that they were dangerous and had to be killed to escape the long hand of the law, thus the emphasis on sexual aggression. The colonialists wove convincing stories of the bloody mating battles of oriental birds and kept up good trade in animals. Darwin of course did not realize this and accepted any account given by colonialists and others invested in such illegal trades *etc.*

Actually this dilemma that troubled Darwin also is a hindrance for any research in natural science even today. And scientists will not have the patience to watch their objects of research by becoming a part of the local niche. No doubt there is a good insight among contemporary scientists that laboratory research may not tell anything worthwhile about animals in wild nature. This may be true for research on domestic cattle also. In ornithological research several methodologies have been developed to do real field observations, such as netting and ringing. Both require environmental clearance and can be very expensive.

But that apart it seems the technology adapted in such practices is not viable. Earlier ringing meant they netted birds in the wild and attached bands made of light alloys with registration numbers. Ringing is a method generally used to study migrating birds. But now-a-day they have started to use electronic chips in the bands which are supposed to continuously emit signals to the nearest microwave tower. But I found that in reality the electronic chips in the bands are so strong that a ringed bird cannot fly very far from any nearby relay station. In the Bhadra area where I did field observation I came across a Forest Eagle owl (Bubo nipalensis) who was banded and couldn't fly away with his flock and is grounded in the region. All he can do is fly in the area around a nearby microwave station. The chip implant in the band is so strong that he cannot overcome the force exerted by the relay tower on it. At this rate I do not see what quality research is being carried out by so called qualified scientists in the field. A couple of years ago the news papers reported on a large number of migrating swallows that committed mass suicide by dashing against the glass panes of a high rise building in Calcutta city. Some scientists even claimed that this revealed that birds were instinctual creatures without much intelligence and could not cope with the changing habitat scenario to find alternative routes for migrating. It is my suggestion that among them the leading adult birds were ringed with electronic chips and the rest of the flock was in the first season and could not be abandoned. The ringed adults could not physically choose a different path which could have avoided that route because a powerful microwave tower was exerting its force on them via the electronic chip in the ring. Such are the feckless ways of the so called ornithological scientist of today.

Reading Darwin, or a Huxley or even Dobzhansky or Mayr, that is up till the early mid century on evolution or species existence is quite a different experience from reading a more contemporary scientists such as say Templeton or Lande or Wright. Not only is there a great deal of advancement made in the direction of genetics there is also a drastic change in methodology from empirical to theoretical modeling and from qualitative to quantitative studies. As can be predicted this has alienated the natural scientist from real world of animals and its laws. A lot of tough looking calculations are being substituted as laws of nature based on wrong premises. I think there is a dire need for specialists in the discipline to go out in the field and get back on a SMS Beagle voyage again.

Because of the advancement in genetics the focus has become increasingly on the microevolutionary processes of speciation, and a revival of Darwinism in the form of synthesis between biological and paleontological study. During Darwin's time the finding of large scale fossil evidence threw a doubt on Darwin's theory of gradual evolution of species but in the twentieth century Darwin saw a reversal of fortune with the development of genetics, at least in theoretical discussions. Genetics actually has remained an inaccurate discipline, with scientist trying frantically to make coherent the contrary evidences thrown up every now and again that raised very fundamental doubts about genetic origins of morphological adaptations and species. Notions such as convergent or divergent adaptations, disruptive or stabilizing selection and local adaptations have been proposed to explain away some of the contradictions. If what I have argued in the book is proved about adaptations and speciation then that entire race for evolution would merely seem like a wild goose chase, which in-resident academics and institutional scientists will not like. Most importantly they will have to think of new mathematical modeling equations especially on population genetics.

Appendix 4

Tree of life as an alternative to Dobzhansky's population genetics model of speciation

In the period between 1937 and 1950, Theodisius Dobzhansky brought a sea change in the study of species and speciation with his book Genetics and the Origin of Species. It brought speciation phenomenon to the research lab (even tray) with its population genetics approach that reduced speciation to intergenerational gene flow under hostile ecological conditions. This reductive view enabled to predict mathematically any speciation event in a population by writing algebraic equations called as predictive modeling. No doubt he meant big business for scientific institutions even if it crowded natural science text books with incomprehensible clever mathematical equations than with real pictures of animals and birds.

In the 1951, third edition of his book he wrote,

Among the different kinds of populations that exist in nature, the organism like integration is most evident in the breeding association which are formed in all sexual and cross fertilizing organism. The integrating agent in such Mendelian populations is the process of reproduction itself, which establishes mating, parenthood, and progeny bonds between the component individuals. A Mendelian population is, then a reproductive community of individuals with share in a common gene pool.

A Mendelian population can be said to possess a corporate genotype. The population genotype is evidently a function of the genetic constitution of the component individuals, just as the health of an individual body is a function of a soundness of its parts. The rules governing the genetic structure of a population are, nevertheless, distinct from those which govern the genetics of individuals, just as rules of sociology are distinct from physiological ones, although the former are in the last instance integrated systems of the latter (Novikoff, 1945). Suppose for example some factors have arisen in the environment that discriminate against too tall or too short individuals of a species. From the standpoint of an individual some growth genes would have acquired lethal properties, and the effects of these genes might be described adequately by stating the precise nature of physiological reactions reaching death. From the viewpoint of population genetics, death of this category of individuals initiates a complex chain of consequences: the relative frequencies of homozygote and heterozygote for certain growth genes and for genes located in the same chromosomes would be altered; some genetic factors which previously were being eliminated because of their harmfulness might become neutral or even favourable; after some generations the genetic constitution of the species may be changed.

Evolution is a change in the genetic composition of populations. The study of mechanism of evolution falls within the province of population genetics. Of course, changes observed in populations may be of a different order of magnitude ranging from those induced in a herd of domestic animals by the introduction of a new sire to phylogenetic changes; leading to the origin of new classes of organisms. The former are obviously trifling in scale compared with the latter. Experience shows however, that there is no way towards understanding of the mechanisms of macro evolutionary changes, which require time on geological scales, other than through understanding of micro evolutionary processes observable within the span of a human lifetime, often controlled by man's will and sometimes reproducible in laboratory experiments.

Many authors believe that micro evolutionary changes are different in principle from macro evolutionary ones, and that while the former can be understood in terms of known genetic agents (mutation, selection, genetic drift) the latter involve forces that are experimentally unknown or only dimly discerned....Well known writers have supposed macro evolutionary changes to be engendered by some directing forces either inherent in the organism itself or acting on it by some inscrutable means from outside.... (Dobzhansky,1951)"

The extract above is a good record of our ignorance about the nature of the gene. In addition perhaps it is in order that a laboratory geneticist should be so totally without any insights into Nature and biological behavior of organisms. My field observation suggests that in case a particular trait should be a drawback in a niche hostile to it, the organism's "some growth genes" would not develop any "lethal properties" as he predicts. In a Mendelian critical population where there is genetic stability the individual would develop novel pleiotropic gene mechanisms to overcome or supplement the handicap rather than lose the trait. In my experience if the handicap leads to death then the population will produce more individuals with the same make up to compensate for the loss. An individual member's experience is inadequate in itself to bring about change in a species stock and the population will remain indifferent to his angst. The gene today is still being defined in positivist terms as carrying unique codes, some of them unique to species and some other unique to individual. Yet when we turn to what we have learnt about it, it is remarkable that much of it is from merely in situ observations of genes in action. It is very likely that the time is ripe to revise our definition and to start viewing the gene as merely an organic template for intelligent actions of organisms rather than carrying the codes themselves. True, there is an element of continuity in the experiences of organisms across generations and individuals, which elicits the existence of a mode inheritance. But the gene has not answered this demand as yet.

Secondly, to come back to Dobzhansky's population genetics, it is important to see that the purpose of a Mendelian population to a species is to ensure its cohesion or to preserve its uniqueness. As I said in the book since speciation is about losing more and more genetic specificity, the proper maintenance of the Mendelian population is very important to species existence. But on the other side, I suggest that the breakup of a Mendelian population and speciation are two different processes involving two different genetic mechanisms. Its break up dissolves the species and makes it to become extinct but this will not lead to the rise of a new species. In wild nature it is true that species show the ability to evolve adaptations suitable to the allopatric conditions that enable them to maximize fitness. In the book I have described many such instances, the magpie robin beak structure or the purple sunbird's or tree pie or racket tailed so on. But it would be wrong to relate these selections to positivist view of genetic change. From my field observations it can be derived that the nature of the Mendelian critical population and the reproductive praxis engaged in by a species is already adapted to allow such selections that maximize niche. They do not need any genetic changes to evolve but on the contrary a genetic ad hoc ness is maintained by the species.

Now of course the question will arise of what is the nature of a Mendelian population? What is the nature of its universal practice among birds? As we know a Mendelian population is made of a critical number of allele (or its multiples which will always a subspecies population, more of this below) members that share a similar genetics among themselves. The members of such a critical population are like the different parts of the same plant, the root(s), flower(s), stem(s), leaf(ves) so on. This analogy is very integral to our understanding of Mendelian genetics and not merely a metaphor. Each allele performs an analogical function as these components in a plant and has comparable genetic makeup. The permanent loss of an allele will kill the plant but not lead to a different plant: Intergenerational gene flow in such a reproductive praxis would mean that each allele of the critical population will take on or move to different positions (root, auxiliary bud, flower, stem, leaf so on) in succeeding generations until they fall off the plant and become extinct. When all the alleles have flown like this and become extinct that species will have become extinct. This is the etiology of gene flow in any species, regardless of individual's experience in it. This suggests that species can come to an end naturally and not by selection or any other factor. In fact the longevity of a species may be totally dependent on the number of alleles in its critical population because this can change from species to species according to their genetic makeup. Mathematically it would work out as the Cpa(n)!= T, where Cp is the number of alleles in the critical population, a(n) number of possible gene antinomies, and T is the life span of the species.

A critical population of a species can exist in multiples. But these multiples will always exist in the form of subspecies (or population alleles). Such subpopulations are like auxiliary buds on the species plant and can give it a new life should the parent plant die off, but they can also exist when the parent plant is alive like branches on a tree. These sub populations also come under the same gene flow pattern as for the parent population. The phenomenon of clines gives good evidence for understanding the nature of subspecies existence and its relation to parent population. In the world of vegetation different plant species have different numbers of auxiliaries that grow on different places on the plant. I suggest this must be true of birds and animals also. Some bird species can have more or fewer than some other.

Further it suggests that there is a relation of reverse homology between intra species genetic dynamics and interspecies. We can predict that death of a species should initiate the birth of a new species at level of clade. This reverse homologic response should also be the relation between the other levels of species existence on this earth. Thus subspecies come into existence by maturing of a species in time; on the other side death of a species will always be followed by progressive extinction of all its subspecies. But when this has been achieved there is bound to origin another new species in the family cluster. Such processes shall continue to be until the family has exhausted its genetic possibilities. By a continued reverse homology when this has happened there should, in turn, come into existence a new family, and so on. Species diversity and longevity is therefore limited only by mathematics. But I calculate that the number is n/o=∞ (where n= no. of genes at any given time in the speciation process, o= number of species specific genes required for speciation) none less than that.

I have said above that new species come into existence by the loss of genetic specificity and by acquiring novel pleiotropic adaptations. This I have said must happen to the karyotype. What are the conditions for this to occur? Drastic or sudden changes in environmental factors are of great importance to bring about these changes. But my field observation points to an intriguing fact that these allopatric factors act only as catalysts in bringing about the change; but inner genetic factors are of more fundamental causes for speciation. When a species has matured sufficiently in time through generations (by inner laws rather than response to allopatry), its members in critical population can get an inner calling to speciate further. This inner genetic dictat then will drive them to seek suitable environmental conditions for catalysts. This process is comparable to an auxiliary bud on a mature plant giving rise to a new branch (in the case of subspecies) or a fruit-seed ripening and falling off the plant in order to sprout a new plant (in the case of a new species); like the spores of a plant floating to find a suitable ground to shoot. For instance in the Bhadra region we have four sub species of leaf birds. It appears that they speciated by undertaking long distance migration from the

Himalayas, but this occurred only at the behest of genetic intuition to diversify. It is possible that if the allopatric conditions are not good the auxiliary bud would fall off the plant without branching; it could lie fallow until suitable time or the parent plant would have to put forth once again, similarly among birds but the urge to speciate cannot be infinitely stalled because it is as natural as growth.

One of the conundrums of evolutionary theory has been the absence of fossil evidence for the gradual development of life forms, and especially for the theory that life evolved from smaller simpler organisms to larger more complex forms. We know that fossil evidences discovered in Britain, North and South America during Darwin's time suggested large geological breaks in earth's natural history. To Darwin it appeared as a blow to his theory of gradual evolution through micro processes of adaptations and natural selection. To his opponents like Aldous Huxley for instance it appeared as proof that evolution was punctuated and took place in sudden spurts and leaps on a large geological time scale.

In this regard in chapter nine of The Origin of Species, Darwin endeavours to present before us the large time scale of our earth's geological history in order to explain the contradictions that arise due to lack of fossil evidence for intermediary forms. Darwin's chapter on geological records appears like a tutorial for children compared to what paleogeologists are claiming today. Today they are in no uncertainty that the Earth's evolutionary history can be divided into three eras as Paleozoic, Mesozoic and Cenozoic to which they trace the emergence of continental land masses and the gradual rise of life forms from small to larger creatures. To a time before these eras are assigned the rise of unicellular organisms. The three eras are divided on the basis of fossil evidence of macro catastrophic events. To these macro ecological events are attributed the extinction of older forms on a large scale. These events are labeled as extinction boundaries because of the strange lack of fossils of the extincted animals. Between the Paleozoic and Mesozoic eras we have the Permian-Triassic extinction event, between the Mesozoic and Cenozoic we have the cretaceous-Paleocene extinction boundary. All these extinction events are marked by lack of fossil evidences. I suggest these boundaries should not be viewed as "extinction" boundaries, especially as created by catastrophes but merely as times when new forms of species were born, not by natural selection, but by the natural coming to end of an earlier form. The catastrophes in fact must have brought about the appropriate allopatric conditions for the formation of new species, albeit because the older had sufficiently advanced to a critically mature genetic makeup so as to enter into a further phase of speciation. The record gaps in the boundaries therefore must be because this was the time new forms came into existence by speciation and no wonder there are no fossil records of them yet. And moreover these geological events in themselves did not bring about extinction of earlier forms but that the forms had come to an end naturally in time like I described above and any of its surviving members (perhaps Darwin's missing intermediary forms) must have undergone speciation in the presence of environmental catalysis.

We know that the modern day birds are called as neo avis. The story of the evolution of neo avis is incomplete and full of challenges. There is enough fossil evidence I believe to prove that there were creatures with wings before the first neo avis archaeopteryx or the vegavis, eliciting the hypothesis that the neo avis evolved from them. The strangest thing is there is no evolutionary evidence that the neo avis developed gradually from these winged creatures. In fact the cretaceous-Paleocene (or the K-T) extinction boundary poses a blank between the primitive winged creatures and the neo avis. Some hypotheses even propose convergent adaptations (of wings and claws) between the two neoavis and primitive winged creatures and not evolutionary link. My hunch is that the neo avis must have speciated from the primitive birds during the K-T event but this first generation neo avis must have faced a crisis and became extinct almost immediately they speciated. The crisis was not any ecological condition but must be internal to the birds themselves. That's why we do not find any fossil evidence for the extinction of these birds and draw a complete blank on evolution of neo avis. What was this crisis that the first generation neo avis faced? Could it have been that they had not developed their display and song mechanisms? This is a puzzle that remains to prove of course.

Unlike some scientists such as Stephen Gould and others it is not that I am denying that the evolutionary processes are always micro processes but only that there is no selection involved in this evolutionary micro process linking specific morphological adaptations to changes in genetic makeup. Speciation and adaptations must be two separate phenomena. It is true that the latter, *i.e.,* acquisition of adaptations has a functional basis (*e.g.,* one can claim that flycatchers have wide chute like mouth to catch flies efficiently, *etc.*), while the former is a relentless process of change best encapsulated by the idea of entropy (a claim that nothing in the universe is static and everything is engaged in a process of decay or degradation). There have been catastrophe theorists such as Norman Sewall who attribute speciation as a macro event, such as those happened at the beginning of earth's geological history. And there have been like Stephen Jay Gould scientists who look at evolution as punctuated equilibria on the basis of these large geologic breaks in fossil record. Then there have been some others who have interpreted macro geological events as having arrested the general progress of species by bringing about allopatric conditions indifferent to them rather than out of the nature of speciation process itself. On the contrary, I suggest that what appear as selection are merely the gaps created by the time-bound death or coming to an end of certain forms by growth in them. But which lead to further speciation by the same virtue. It is possible newly emerging species, especially the large sized creatures of the beginning periods of evolution (like I believe evolution to be), brought about the allopatric conditions for their own speciation process, and may have caused the geological events. I suggest that evolution must have been a far more exciting dialogic event than a simplistic development from small, simple to large complex genetic forms fighting against their environment as well as among themselves.

Thus, even though species once come into existence by the natural processes of growth and maturation must perforce become extinct in time by the same, even under favourable allopatric conditions. Strongly unfavourable ecology cannot in the last instance hinder the progress of evolution. Nor can it bring about extinction of a species before its time. One, it is possible that there will be a decline in population numbers to the extent of ecological handicap. A second possibility is that species may hold in abeyance its own development by going underground (latent or dormant) to suit environmental (including other animals) demands. It is not that I am indifferent to contemporary environmentalist angst and eco activism; perhaps it may be reminded that habitat degradation by humans can still have an impact on species numbers negatively to human well being. And lovely species can deny us their benefit until we are extinct from its habitat.

Appendix 5

Some speculations on the originary speciation event

The origin of life forms is a topic that has held human imagination from the earliest ages. The diverse cultures and their religions in the entire world have envisioned different narratives about the creation of life on this earth. Most of such thinking has at its centre the idea of a god and goddess as the originator. Until the discovery of an immense data of fossil evidence in the nineteenth century and the formulation of an evolutionary theory by scientists such as Charles Darwin, challenged the older cultural religious narratives. Ever since there have been two contesting camps creationists and evolutionists.

Any theory of speciation is bound to entangle with a conception of some sort of originary event now or later. Such an engagement, I have begun to feel in my exploration of birds of the Bhadra region and the purpose of song, is bound to lead one to confront the secular with the non secular rather than discard any one. By conviction I have always held on strongly to a radical atheism but this experience of nature leads me to consider that perhaps religious attribution of origin of life to a god must be true after all. And below I present a hypothetical story of an originary event of life on this earth on the basis of insights I have gained into genetics, species, birds and songs.

I have said that speciation events must be looked as involving loss of specificity of genes for the purpose of acquiring a novel pleiotropic gene expression. A new evolved organism is always genetically simpler, has fewer numbers of specific genes but acquires an increasing number of complex novel gene expressions, or in other words pleiotropy, that are co adapted on the lost gene. In organisms of today, strictly speaking what is glamorized as genes is nothing but a quantity of degraded (may be semi digested fatty foods) fatty molecules located at the center of the biological cell. All poly acids and proteins have the double helix structure that DNA molecule has. But what is impressive is that such fatty molecules at the centre of the biological cell serve as templates for the intelligent functions of the organism. Still it is important to see that these intelligent functions are more in situ observed and localized phenomenon and very difficult to assign to the specificity of these fatty molecules. Research on gene functions throws doubt on the very definition of DNA as genetic encoding and suggests the possibility that it is merely organic material that enables to provide typical responses to stimuluses.

If we imagine backward in the history of the earth on an evolutionary time scale this would mean at some stage in Earth's history there must have lived a first originary organism from which speciation is derived. This first originary organism was genetically far more complex than even the highest order animal of today, but it was so specific that its gene functions were clearly specified and therefore simpler, *i.e.,* it had a billions of different gene for the billions of tasks/functions, the shape of which unimaginable gene mass manifesting as an infinite-cellular organism with as many eye-cells with which to see the 380 degrees, and as many ear-cells and as many limb-like cells, so on. Such genetic complexity must have made it god like, super intelligent, all knowing and omniscient. At one point in its life time must have thought of the future and perpetuating for posterity. Speciation was the mechanism adopted by this originary organism. Speciation is a primary process by which nature reproduces itself.

The originary organism must have been an auto generating one which could have reproduced itself again and again and continue to retain its specificity. But superior as it was genetically it must have in a flash recognized that in it there was an infinite possibility of life which would not be realized should it perpetuate. The originary speciation must have been an act of supreme altruism or a sentimental act because by it the originary organism annihilated its own specificity so that we the descendents may survive in it. The first pleiotropic gene expression must have been that of 'sentiment' located on a gene that in the originary organism gave its uniqueness or wholeness. As I said above a gene is never completely deleted or lost and always survives as a negative value. Thus the site of the deleted specific gene became the source of sentiment in the form of a novel pleiotropic epistasis, whose nature was attachment or sentiment handed down to subsequent generations of species by the originary ancestor. In the creatures born at the beginning it appeared as root as in plants and in embryo, but as speciation proceeded it took on even stronger novel pleiotropy and became source of filial love and perhaps basis for song and communication.

The way the originary organism conceived of speciation safeguarded the perpetuation of its descendents by diversity. And all this rested on that one act of loss of specificity of its own uniqueness. Then this organism must have been no less than god. But since then, by it we are handed down with speciation as a ritual of negation rather than a positive bequest of genes. As I said the universe of life on the earth is a perpetual race against genetic predeterminism and is heading towards chaos or anarchy. The collateral is of course creativity, freedom of gene expression and connectedness of beings.

Perhaps when we have speciated infinitely (n/o=∞ where n =no of genes of the originary organism, o=number of genes required for new species to form) we will have reached a state of utter decadence, when the initial bequest of sentiment/attachment will have attained its antinomy in the expression of hatred and war then an act of reversal stemming out of our selfishness will take us all back to this originary god. Perhaps this is the act which many religious texts refer to as Kalki avatr (Hindu) or second coming (Christianity). I am speculating of course and perhaps we will find no fossil evidence for the originary speciation because the first form did not die but sublimated itself to produce the first species.

A second and most significant evolution that occurred must have been the development of the brain. The originary god did not have brains but only genes. But in successive speciation events as specific genes were lost and replaced by more and more pleiotropy mechanisms that coordinated the genes that survived, it gave rise to the brain. Actually a homology may be found between such a speciation process and the way organisms are born at all levels of the tree of life. Down the evolutionary tree we find organisms with less brain and more specifically adapted appendages and complicated body (for example in primitive algae and ferns), while later evolved organisms have larger brain to coordinate the simplified morphology.

Even though as I said it may be never proved that life on this earth originated the way, because the ever evolving life forms are but the living fossil evidences testifying to the originary creator's existence. Anyway there is ample fossil data to scandalize the theory of evolution proposed by Darwinists from several directions. The theory of evolution seems to rest merely on narrative evidence (way of telling the story) rather than on concrete facts. Take for instance the most recent story of assignation of Burgess's Shale fossil fauna uncovered in the Canada Rockies in the year 1909. The American paleontologist Charles Doolittle Walcott in 1909 refused to commit himself to the early Cambrian explosion of life theory, because it would upset Darwin's theory of gradual evolution of life. He couldn't trace the Burgess explosion of life to any pre-Cambrian macro-speciation event either and instead placed in as very very recent formation in the Cenozoic even Pleistocene era, *i.e.,* close to our time. Successive paleontologists after Walcott starting with Charles Knight Kinsley have not seen any reason to seriously challenge his views. Stephen Jay Gould's defense of the Cambrian explosion theory in his book Wonderful Life even though most eloquent and dramatic but does not succeed further than that his colleagues Henry Whittington, Derek Briggs and Simon Conway Morris starting 1971 had found intermediary forms for our modern day marine life that fitted Darwin's theory of speciation like kid glove, and are forced to downplay it. Similarly, Gould chooses not to address the question of nature of fossilization of the Burgess biota. Much of the new world fossil finding is either chalk, or limestone, or greensand and therefore can but come from a recent past: Gould does not give a clue on this subject.

It seems to me that the desperation to show that the tiny marine creatures of the Burgess Shale came from a far earlier period is because any modern dating challenges the small to big and, simple to complex evolution narrative, especially on Gould's part as appears by his engagement with the creationist challenges (example in his book The Panda's Thumb). This – small to big, simple to complex - is the unacknowledged theoretical premise of both camps Walcott and Gould, and which does not have enough proof in fossils. I suggest that there is something naïve in paleontologists' search for microscopic protozoan (they have found enough in the pre Cambrian layers by now) that will explain the origin of multicellular organisms. It seems likely that multicellular organisms of today branched from a multicellular descendent of the originary creature far far larger. The protozoan microbes we find today must be only one line of the originary creatures' descendents, and more recent than the multicellular life forms of today.

If we pursued the possibility of a god-like first organism like I proposed, and the setting in motion of a great chain of speciation events, then the first life forms must have been indeed immensely huge. It is possible that in the beginning speciation event proceeded by giving rise to large algae like plants that spread across the oceans, then there must have been a period of hybrids neither plant nor animal beings, neither unicellular nor multicellular but a cellular biota complex (as the tomatians have been reconstructed by some scientists), then the large sized ontologically individualized water animals came to exist and perhaps simultaneously came the amphibian reptiles and birds, lastly the mammoth land animals so on. Life got smaller and smaller as speciation advanced in each of these lines, and only in the modern age were the tiniest creatures formed.

Gould in his book justifies the Cambrian explosion theory by showing that many of these novel creatures of the Burgess Shale lost the trial for survival and became extinct rather than assign them as intermediary forms that lead to modern phyla. His theory of Survival of Fittest shows how out of touch he is with wonderful life and suspiciously smells of masculism. To support his theory of survival of the fittest he gives the examples of two species of extinct flightless land birds, one diatrymids that lived on the Northern American and European continents after the cretaceous and a second phororhacid that existed in a later period on the South American continent. Both were carnivore and preyed on mammals. He views the extinction of the first to the dominance of mammals on the North and European continent and survival of the second to the absence of mammals on the South American continent. I suggest that contrary to Gould's and A.S Romer (on whom he draws) the primitive predator diatrymid must have been a key species that aided in the evolution of mammals as purely and strongly land animals: *i.e.,* for their distinguishing features from other water and air creatures dominant at the time, speedy locomotion on land. I suggest that the diatrymid birds had weak legs but strong neck and head suggesting that they were not really hunting but parasiting creatures somewhat like hyena and wolf. The diatrymid must have found food on the vast savannah and steppe plains of the two continents by hustling the hunting mammals younger in form than they such as lions and tigers but with strong legs. Both preyed on smaller mammals. The presence of the diatrymid older in form and dependent on the younger mammals for providing food must have directed the course of evolution of mammals to become powerful on their legs and totally land animals. I see a moral design in natural evolution in that the diatrymid could have evolved to become an ostrich or a Kiwi stronger and powerful on legs than even the fastest mammals but then mammals would not have evolved on the continents. When these land birds had finished their mission they diversified by speciation further not unlike the mammals their protégées so far but now come of age and did not need their help any further. The second generation of flightless birds such as phororhacid comparable to the diatrymids but on the Southern American continent also present a similar picture. I suggest that the carnivore marsupials predominant on the continent had weak legs to hunt the faster smaller mammals such as armidillos and needed the help of the phororhacids that were with stronger legs but weaker neck. The two must have evolved together on the cutoff continent and developed complimentary adaptations to suit each other. Diversity and law of compensation seem to be the rule of nature rather than survival of the fittest. Whatever it is the mystery of life is yet to be solved regardless of Darwin and evolution and nothing is as cut and dried as scientists would like to make out.

Bibliography

Field Guides

Ali, Salim(2002),The Books of Indian Birds (Thirteenth Edition), BNHS & Oxford University Press, New Delhi & Ripley, Dillon S. (2001), Handbook of the Birds of India and Pakistan Volumes 1 to 10 (Second Edition) BNHS & Oxford Universtiy Press, New Delhi.

Baggaley, Ann & Leon Gray (2009), Do You Know the most Speedy, Greedy, Noisy, Birds...in the World?, Bounty Books, UK.

Erritzoe Johannes, Clive E. Mann, Fredrick P. Brammer & Richard Fuller (2012), Cuckoos of the World Christopher Helm, UK.

Ghorpade M Y. (2005), Winged Friends Karnataka Seva Sangha, Karnataka.

Grewal, Bikram(2000),Birds of the Indian Subcontinent (third Edition) Local Colour, Hong Kong.

Grimmett, Richard & Inkskipp, Tim (2007),Birds of Southern India Om Field Guides, New Delhi.

Harrison, Collin & Greensmith, Alan (1993), Birds of the World Dorling Kindersley Lmt, Great Britain.

Hume Julian P & Walters Micheal (2012), Extinct Birds T & AD Poyser, UK.

Kazmierzak Krys & Ber Van Perlo (2000), Birds of India, Sri Lanka, Pakistan, Nepal, Bhutan, Bangladesh and the Maldives Om Field Guides, New Delhi.

McArthy E.M.(2006),Handbook of Avian Hybrids of the World, Oxford University Press, UK.

Moss, Stephen(2007),Collins Remarkable Birds Harper Collins, UK.

Narsihman, Dr. S V.(2004, 2008), Feathered Jewels of Coorg (Second Edition), Coorg Wildlife Society, Madikeri.

Singh, Rajpal (2005), Birds of Bharatpur, Prakash Books, New Delhi.

Tejaswi Purna Chandra K P (2010), Mayeya Mukhagalu Pusthaka Prakashana Karnataka.

Vriends, Matthew M. & Heming-Vriends, Tanya M. (2004), The Handbook of Cage and Aviary Birds Silverdale Books, UK.

Willem den Besten, Jan (2008), Birds of India and the Indian Subcontinent, Mosaic Books, New Delhi.

Other References

Ali, Salim (1985), The Fall of a Sparrow Oxford University Press, India.

Alcock John (2005), Animal Behaviour; An Evolutionary Approach (Eight edition Sinauer Associates, USA.

Armstrong, E A., (1973), Study of Birdsong, Dover Press, New York.

Arnold, David & Guha, Ramachandra (Eds) (1995, 1998), Nature Culture and Imperialism (Second Edition) Oxford University Press, New Delhi.

Bolhuis, John J. & Martin Everaert (Eds) (2012), Birdsong, Speech. And Language: Exploring the Evolution of Mind and Brain (with a foreword by Robert C Berwick & Noam Chomsky) OUP, UK.

Brown, J. L. (1975), The Evolution of Behaviour W W Norton, New York, (1987) Helping and Communal Breeding in Birds: Ecology and Evolution, Princeton University Press, Princeton.

Burton, Robert (1985), Bird Behaviour, Granada, UK.

Calow P (Ed) (1998), Encyclopedia of Ecology and Environmental Management, Blackwell Press, UK.

Carson, Rachel (2000), Silent Spring, Penguin Modern Classics, UK,

Catchpole, C. K. & P. S. B. Slater (1995), Bird Song: Biological Themes and Variations, Oxford University Press, U.K.

Collias, Nicholas E. & Collias Elsie C. (1984), Nest Building and Bird Behaviour. Princeton University Press, USA.

Darwin, Charles (1859, 2008), On the Origin of Species, ebook Gutenberg Project, (1860 ,2008), A Naturalist Voyage Round the World, ebook Gutenberg Project. (1867 ,2008., 2012), The Descent of Man, ebook Adelaide university. (1899, 2012) The Expression of the Emotions in Man and Animals, ebook Gutenberg Project, (1909,2007), The Foundations of the Origin of Species, ebook Gutenberg Project.

Davies, Nicholas B Krebs John R & West Stuart A.(1981),An Introduction to Behavioural Ecology (Fourth Edition),Wiley Blackwell, UK.

Dawkins, Richard (1976), The Selfish Gene, Oxford University Press, New York. (1986) The Blind Watchmaker, W W Norton & Company, New York.

Dobzhansky, T. (1951, 1982), Genetics and the Origin of Species (with an introduction by Stephen Jay Gould), Columbia University Press, U S A.

Ferguson, J W H. (1999), "The significance of mate selection and mate recognition in speciation" In Proceeding of the 22 International Ornithology Congress, Durban edited by Adams, N J., Slotow , R. H.) Birdlife, Johannesberg, South Africa, pp.1496-1504.

Fuller, Robert J. (Ed) (2012), Birds and Habitat; Relationships in Changing Landscape, Cambridge University Press, UK.

Gooders John (1992), The Survival World of Birds, Boxtree, UK.

Gould, Stephen J. (1988/89), Wonderful Life: The Burgess Shale and the Nature of History, Hutchinson Radius, UK.

Greenwalt, C. H.(1968), Bird Song: Acoustics and Physiology, Washington Smithsonian Institute Press, USA.

Habib, Irfan (2010), Man and Environment The Ecological Survey of India, Tulika, India

Halliday, Tim (1978, 1980), Vanishing Birds Their Natural History and Conservation Penguin Books, UK. (1980), Sexual Strategy Oxford University Press, UK.

Hamilton, W. D. (1975) "Innate social aptitudes of man, an approach from evolutionary genetics" in Biosocial Anthropology, edited by R. Fox), pp.135-155, John Wiley & Sons.

Hansk, Ikka A. & Gaggiotti Oscar E. (Eds.) (2012), Ecology, Genetics, and Evolution of Metapopulations, Oxford University Press, UK.

Jameson Conor Mark, (2012), Silent Spring Revisited, Bloomsbury, UK.

Jinks, J. L. (1983), Biometrical genetics of heterosis (Heterosis: Reappraisal of Theory and Practice edited by R Frankel), Springer-Verlag, Berlin, pp.1-46,

Komarek, Stanislav (1998), Mimicry, Aposematism and Related Phenomena in Animals and Plants, Prague Publishers, Vesmir.

Kroodsma, D. E. (1982), Song repertoires: Problems in their definition and use (Acoustic Communication in Birds edited by Kroodsma D. E. & Miller E. H.), Academic Press, New York pp.125-146.

Lorenz, Konrad (1949), King Solomon's Ring, USA.

Lack, David (1971), Ecological Isolation in Birds, Blackwell Scientific Publications, UK.

Leisler Bernd & Schulz- Hagen,Karl (2012),The Reed Warblers: Diversity in a Uniform Bird Family. KNNV Publishing, Netherlands.

Levin A. Simon (Ed- in- chief) (2001, 2007), Encyclopedea of Biodiversity Vol 5, Academic Press, UK.

Mayr, Ernst (1942, 1999), Systematics and the Origin of Species; From the view of a Zoologist Harvard University Press, U S A.

Mies, Maria & Vandana Shiva (1993, 2000) Ecofeminsm Reader, Basil Blackwell Press, UK.

Morris, Desmond (1967, 1994), The Naked Ape ,Vintage edition, UK. (1969, 1994), The Human Zoo, Vintage, UK, (2004, 2005), The Naked Woman, Vintage, UK.

Nowak, Martin A. & Highfield, Roger (2011), Super Cooperations: Altruism, Evolution, and Why We Need Each Other to Succeed, Free Press, New York.

Nosil, Patrick (2012.), Ecological Speciation, Oxford University. UK,

Owen, Dennis (1980), Camouflage and Mimicry, Oxford University Press, UK.

Parry, James (2012), Mating Lives of Birds, New Holland Publishers, UK.

Paterson, H. E. H.(1985), "The recognition concept of species" (In Species and Speciation, edited by E. S.Vrba), Transvaal Musuem, Pretoria, pp.21-29. (1993) "Variation and the specific mate recognition system" (In Perspectives in Ethology, Vol 10, edited by P. P. G. Bateson, *et al.,*) Plenium Press, New York, pp.209-227.

Peterson, A. Townsend (2012), Ecological Niches and Geographical Distribution (monograph), Oxford University Press , UK.

Poulton, Edward Bagnall (1890), The Colours of Animals, their Meaning and Use, Especially Considered in the Case of Insects, Kegan Paul, Trubner & Co, London.

Rand Austin L. (1964), Ornithology: An Introduction, Penguin Books,UK.

Renner Swen C & Rappole John H (Eds) (2011), Avifauna of the Eastern and Southeastern sub-Himalyn Mountains: Centre of Endemism or Many Species in Marginal Habitats?, American Ornithologist's Union, USA.

Ruxton, E. B., T. N. Sherratt, & M. P. Speed (2004), Avoiding attack: the Evolutionary Ecology of Crypsis, Aposematism and Mimicry, Oxford University Press, Oxford.

Searcy, William & Stephen Nowicki (2005), The Evolution of Animal Communication, Princeton University Press, Princeton.

Sepkoski, David (2012), Rereading the Fossil Record: The Growth of Paleobiology as an Evolutionaray Discipline, University of Chicago, USA.

Shiva, Vandana (1989, 2005), Staying Alive; Women, Ecology and Survival Kali fopr Women, New Delhi.

Smith , Maynard & G. R, Price (2003), Animal Signals, Oxford University Press, Oxford.

Stoddard, P. K. (1996), Vocal recognition of territorial passerines (Ecology and Evolution of Acoustic Communication in Birds edited by Kroodsma D. E. & Miller E. H.), Cornell University Press, Ithaca New York, pp.356-374.

Stott, Rebecca (2012), Darwins Ghosts In search of the First Evolutionist, Bloomsbury, UK.

Trivers, R L (2002),Natural Selection and Social Theory, Oxford University Press, New York.

Urfi Abdul Jamil, (Ed.) (2008), Birds of India A Literary Anthology, Oxford University Press, New Delhi.

Van Staaden, Moira J, William Searcy & Roger T Hamiltion 'Signaling Aggression' in Robert Huber, Danika L Bannasch & Patricia Brennan (Eds.) Advances in Genetics, Vol 75, Academic Press, Burlington, 2011, pp.23-49.

Von Neumann, J., & O. Morgenstern. (1944), Theory of Games and Economic Behaviour, Princeton University Press, Princeton.

Wayne-Edwards, V C (1962), Animal Dispersion in Relation to Social Behaviour, Oliver and Boyd, Edinburg.

Watson, James D. (2012), Double Helix (edited by Gann Alexander & Witkowiski,Jan), Simon &Schuster, New York.

Whitacre, David F. & Jenny, Peter J., (Ed) (2012), Neotropical Birds pf Prey: Biology and Ecology of a Forest Raptor Community , Cornell University Press, USA.

Whittaker, Robert J. & Fernandez-Palacios, Jose Maria (2012), Island Biogeography; Ecology, Evolution, and Conservation (second Edition), Oxford University Press, UK.

William, G.C .(1966), Adaptation and Natural Selection, Princeton University Press, Princeton.

Wilson, E. O. (1971), The Insect Societies, Harvard University Press, Cambridge MA. (1975) Sociobiology: The New Synthesis, Harvard University Press, Cambridge MA.

Zahavi, Amotz & Avishag Zahavi (1997), The Handicap Principle, A Missing Piece of Darwin's Puzzle, Oxford University Press, New York.

Ziegler, H. Philip & Marler, Peter (2012), Nueroscience of Birdsong, Oxford University Press, UK.

Journal Articles

Ayala, Francisco J.(1972), 'Competition between species: The diversity of environments in which most organisms live permits the coexistence of many species, even when they compete for the same resources' in American Scientist, Vol. 60, No. 3, May-June, pp. 348-357.

Axelrod, R & W D Hamilton (1981) " The evolution of cooperation" in Sciences new series Vol 211, No. 4489, pp.1390-1396.

Baker, Myron Charles (1975), 'Song Dialects and Genetic differences in white crowned sparrows (Zonotrichia leucophrya)' in Evolution, Vol 29 No 2 pp 226-241. & J. Cunningham (1985), "The biology of birdsong dialect" in Behavioural Brain Science, vol 8, pp.85-133.

Ballentine, Barbara, Jeremy Hyman, & Stephen Nowicki (2004), "Performance influences female response to male bird song: an experimental test " in Behavioural Ecology Vol 15, pp.163-168.

Baptiste L F, & Robin A Kiester (2005), 'Why birdsong is sometime like music' Perspectives in Biology and Medicine Vol 48, No 3summer pp426-443 & L. Petrinovich (1986) "Song development in white crowned sparrow: social factors and sex differences' Animal Behaviour Vol 34, pp 1359-1371. & P. W. Trail (1992) "The role of song in the evolution of passerine diversity" in Systematic BiologyVol 41 No. 2, pp. 116-125.

Bateson, Patrick (1978), 'Sexual imprinting and optimal out breeding' Letters to Nature in Nature Vol. 273, pp.659-660.

Baugh, Alexander J Kim L Hoke & Micheal J Ryan (2012), 'Development of communication behaviour receiver ontogeny in tungara frogs and a prospectus for a behavioural evolutionary development' The Scientific World Vol , Article ID 680632, pp1-10.

Beecher M.D., P. K. Stoddard, S. E. Campbell & C. L .Horning (1996), "Repertoire matching between neighboring song sparrows" in Animal Behaviour Vol 51, pp.917-923.

Beecher M. D. & J. M. Burt (2004). "The role of social interaction in birdsong learning" in Current Direction in Psychological Science Vol 13, pp.224-224.

Beecher M. D. P. K. Stoddard, S. E. Campbell & C. L .Horning (2000), "Song type matching between neighboring song sparrows" in Animal Behaviour, Vol 59, pp.21-27

Berwick, R. C., Kazuo Okanoya, Gabriel J L Beckers & Johan Bolhuis (2011), "Songs to syntax: the linguistics of birdsong" in Trends in Cognitive science Vol 15, no. 3, pp.113-121.

Brenowitz, E A. (1991) "Evolution of the song control system in the avian brain' in Seminars in the Neurosciences Vol3, pp.399-407.

Brenowitz, E A., D. Margoliash, & K. W. Nordeen (1997), " An introduction to birdsong and the avian song system" in Journal of Neurobiology Vol 33, pp 495-500.

Burley, Nancy Tyler (2006), "An eye for detail: Selective sexual imprinting in zebra finches" in Evolution, pp.1076-1085.

Carson Hampton L. & Alan Templeton (1984), "Genetic revolutions in relation to speciation phenomena; the founding of new populations" in Annual Review of Ecology and Systematics, Vol 15, pp. 97-131.

Chandler, C Ray. Mark H. Gromko (1989), "On the relationship between species concept and speciation processes" Systematic Zoology, Vol 38 No. 2 June, pp. 116-125.

Chilton, G M., M Ross Lein, & Luis F Baptiste (1990), "Mate choice by female white crowned sparrows in a mixed dialect population" in Behavioural Ecology and Sociobiology Vol 27, No. 3, pp. 223-227.

Danner Julie E. Raymond M Danner, Francis M Bonier, *et al.*, (2011) " Female, but not male, tropical sparrows respond more strongly to the local song dialect: implications for population divergence." in The American Naturalist, Vol. 178, No. 1 (July), pp. 53-63.

De Voogd, J J (1991) "Endocrine modulation of the development and adult function of the avian song system" in Psycho neuro endocrinology, Vol 16, pp. 41-66.

Dobzhansky, Theodosius & Olga Pavlovsky (1957), "An experimental study of interaction between genetic drift and natural selection'" in Evolution, Vol . 11, No.3, , pp. 311-319

Futuyama, Douglas J, Gregory C., & E. Mayr (1980), "Non allopatric speciation in animals" Systematic Zoology, Vol 29, No. 3 ,pp. 254-271.

Gould, Stephen Jay (1981, 1994), "Evolution as fact and theory," in Discover 2, May, pp 34-37; Reprinted here with permission from Hen's Teeth and Horse's Toes, New York: W. W. Norton & Company, pp. 253-262. (1980, 1994), "The return of hopeful monsters," in Natural History 86 (June/July)pp 22-30; Reprinted here with permission from The Panda's Thumb New York: W. W. Norton & Co., pp. 186-193.

Gurney, M. E. & M. Konishi (1980), "Hormone induced sexual differentiation of brain and behavior in zebra finches" in Science, Vol. 208, pp1380-1383.

Hamilton, W. D. (1964), "The genetical evolution of social behavior I & II" in Journal of Theoretical Biology, Vol 7, pp.1-52. (1972) "Altruism and related phenomena, mainly in social insects" in Annual Review of Ecological Systems, Vol 3, pp.193-232. & Marelene Zuk (1982), "Heritable true fitness and bright birds: a role for parasites?" in Sciences new series Vol 218, No. 4570, pp.384-387.

Handford, P. & F. Nottebohm (1976), "Allozymic and morphological variation in population samples of rufous –collard sparrow Zonotrichia capensis in relation to vocal dialects" in Evolution Vol 30, No. 4, pp. 802-817.

Herrel, A. J. Podos, S. K. Huber & A. P. Hendry (2005), "Bite Performance and Morphology in a Population of Darwin's Finches: Implications for the Evolution of Beak Shape" in : Functional Ecology,Vol. 19, No. 1, pp. 43-48

Holloway, S. C. & D. F. Clayton (2001), "Estrogen synthesis in the male brainb triggers development of avian song control pathway" in Journal of Nueroscience, Vol 4, pp.170-173.

Hunter M. L. & J. R. Krebs (1979), "Geographic variations in the songs of the great tit (Parus major) in relation to ecological factors" in Journal of Animal Ecology, Vol 48, pp.759-785.

Jarvis, E., D. S. Ribeiro, M. L. Da Silva, J. Viellard & C. V. Mello (2000), "Behaviourally driven gene expression reveals song nuclei in hummingbird brains" in Nature Vol. 406, pp. 628-632.

Jin H. & D. F. Clayton (1997), "Localized change in immediate early gne regulation during sensory and motor learning in zebra finches" in Neuron Vol. 19, pp.1049-1059.

Johnson, S. D. (2009), "Darwin's legacy in South African evolutionary biology" Review Article in South African Journal of Science Vol 105, pp.403-409.

Kirn, A. P. & J. J. De Voogd (1999), "The genesis and death of vocal control neurons during sexual differentiation in the zebra finches" in Journal of Nueroscience, Vol. 9, pp.3176-3187.

Krebs J. R., R Ashcroft & M Webber (1978), "Song matching in the great tit *Parus major L*" in Animal Behaviour, vol. 29, pp.918-923.

Krebs J. R., R. Ashcroft & M. Webber (1978) "Song repertoire and territory defense" in Nature vol 217, pp.539-542 & D. E.Kroodsma (1980) "Repertoires and geographical variation in birdsong" in Advances in the Study of Behaviour, vol 11, pp.113-177.

Kroodsma, D. E., & B. E. Byes (1991) "The function of bird song" in American Journal of Zoology, vol 31, pp.318-328, W. C. Liu, E. Goodwin & P. A. Bedell (1999a) "The ecology of song improvisation as illustrated by North American sedge wren" in Auk Vol 116, pp.373-386, J. Sanchez, D. W. Stemple, E. Goodwin, M.L. de Silva & J.M .E. Vielland 1999b) "Sedentary life style of neotropical sedge wren promote song imitation" in Animal Behaviour Vol 57, pp.855-863.

Kroodsma, D. E. Bruce E Byers, Sylvia L Halkins *et al.*, (1999c) "Geographic variation in black- capped chickadee songs and singing behavior" in The Auk, Vol 116, No 2, pp.387-402.

Lambert, B .M. & H. E. H. Paterson (communicated by T. G. Vallanci) (1983, 1985), "On 'Bridging the gap between race and species':Thw isolation concept and an alternative" in Proceedings of the Lien Society of the N S W, Vol 107, no. 4, pp.501-514.

Lande, Russelll (1981), "Models of speciation by sexual selection on polygenic traits" in Proceeding of the National Academy of Science Vol. 78, No. 6, pp.3721-372. & Steven J. Arnold (1983) "The Measurement of selection on correlated characters" in Evolution Vol 37, No. 6. , pp.1210-1226.

Lemon, R. E. (1975) "How birds develop song dialects" in The Condor Vol 77, No. 4, pp.385-406.

Marler, Peter (1957) " Specific distinctiveness in communication signals of birds" in Behaviour Vol 11, pp.13-39. (1967) "Animal communication signals" in Science, New Series Vol 157, No 3790 pp769-774. (1997) "Three Models of Song Learning: Evidence from Behavior" in Journal of Neurobiology Vol 33, pp 501–516. & M Tamura (1962) "Song "dialects" in three populations of white crowned sparrows in The Condor Vol 64, No 5, pp368-377. (1964) "Culturally transmitted patterns of vocal behaviour in sparrows" Science Vol 146, no. 3650, pp.1483-1486.

Mayr, Ernst,(1982), "Speciation and macroevolution" in Evolution, Vol. 36, No. 6 ,Nov., pp. 1119-1132 (1992), "A local flora and the biological species concept" in American Journal of Botany , Vol. 79, No. 2, Feb., pp. 222-238 (1982), "Of what use are subspecies?" in The Auk, Vol. 99, No. 3, Jul., pp. 593-595. (1956), "Geographical character gradients and climatic adaptation" in Evolution, Vol. 10, No. 1,Mar., pp. 105-10. (1978)," Modes of Speciation, M. J D White" (Review article) in Systematic

Zoology Vol. 27 no. 4, pp.478-482. & Charles Vaurie (1948), "Evolution in the family dicruridae (Birds)" in Evolution, Vol. 2, No. 3, Sep., pp. 238-265.

Mello, C V & S. Ribeiro (1978), "ZENK protein regulation by song in the brain of song birds" in Journal of Comparative Neurobiology Vol 389, pp. 426-438.

Mitton J B & M C Grant (1984), "Associations among protein heterozygosity, Growth rate and developmental homeostasis" in Annual Review of Ecology and Systematics Vol 15, pp.479-499.

Mooney, R.. W. Hoese, S. S. Nowicki (2001), "Auditory representation of the vocal repertoire in a song bird with multiple song types" in Proceedings of the National Academy of Sciences, Vol 98, pp.12778-12783

Nelson, D. A. (1999), "Ecological influence on vocal development in the white crowned sparrow" in Animal Behaviour vol 58, pp 21-36. (2000), "Song overproduction, selective attrition and song dialect in the white crowned sparrow" in Animal Behaviour vol 60, pp.887-998.

Nottebohm F (1969), "The song of the chingolo (*Zonotrichia capensis*) in Argentina: description and evaluation of a system of dialects" in Condor Vol 71, pp.299-315.

Nordby, J. C, S. E. Campbell & M. D. Beecher (1999), "Ecological correlates of song learning in song sparrow" in *Behavioural Ecology* vol 10, pp.287-297.

Nowicki, S (1987) "Vocal tract resonances in oscine bird sound production; evidence from bird songs in helium atmosphere" in Nature, 1987, 325, pp 53-55. & R R Capranica (1986) "Bilateral syringeal interaction in the production of an oscine bird sound" in Science vol. 231, pp11297-1299. & R R Capranica (1986), "Bilateral syringeal interaction in the phonation of song bird" in Journal of Neuroscience, Vol 6, pp3595-3610. & Peter Marler (1988) "How do bird sing?' in Music Perception: An Interdisciplinary Journal, Vol 5, no 4, Biological Studies of Musical Processing, pp391-426, W A Searcy & M Hughes (1998) "The territory defense function in song in song sparrow; a test with speaker occupation design" in Behaviour , vol 135, pp615-618, William A Searcy & Susan Peters (2002) "Quality of song learning affects females' response to male bird song" in Proceedings of the Royal Society of London B 269:no. 1503, 1949-1954 & W. A. Searcy (2005) "Song and mate choice in birds: How the development of behavior helps us understand functions" in The Auk, Vol 122, No 1, pp1-14.

Paterson, H. E. H. (1981), "The continuing search for the unknown and the unknowab;le: a critique on contemporary ideas on speciation, in South African Journal of Science, Vol. 776, pp.113-119.

Payne, R. .B (1978), "Microgeographic variation in songs of splendid sunbirds Nectarina coccinigater populations phonetics, habitats, and song dialects" in Behaviour, Vol. 65, No ¾ , pp. 282-308.

Pimentel, David, G. J. C. Smith & Joyce Soans (1967), "A population model of sympatric speciation" in The American Naturalist, Nov –Dec, pp.493-504.

Podos, J., Susan Peters, Tamia Rudnicky, Peter Marler & S Nowicki (1992) "The organization of song repertoire in song sparrows: Themes and variations" in Ethology Vol 90, pp89-106, Sarah K Huber & Benjamin Taft (2004), "Bird song: The interface of evolution and mechanism" in Annual Review of Ecology, Evolution and Systematics Vol. 35, pp.5-87.

Queller, D. C. (1994), "Genetic relatedness in viscous population" in Journal of Evolutionary Ecology, Vol. 8, pp.70-73.

Ryan, Michael J., William Rand, Peter L Hurd, Steve M Phelps & A Stanley Rand (2003), "Generalization in response to mate recognition signals" in The American Naturalist, Vol 161, no 3, pp 380-394.

Rose, G J., F. Goller, H. J. Gritton, *et al.*, (2004), "Species typical songs in white crowned sparrows tutored with only phrase pairs" in Nature vol 432, no. 7018, pp.753-758.

Salzen, Eric A (1967), "Imprinting in birds and primates" in Behaviour Vol 28, No. 24, pp. 232-254

Searcy, William A. & Malte Anderson (1986) "Sexual selection and evolution of song" in Annual Review of Ecology and Systematics, Vol. 17, pp. 507-533 (1989) "Function of male courtship vocalizations in red-winged blackbirds" in Behavioral Ecology and Sociobiology, Vol. 24, No. 5, pp. 325- 331, S. Nowicki & S. Peters (1999), "Song types as fundamental units in vocal repertoires" in Animal Behaviour Vol 58, pp.37-44.

Slater, P J B (1989) "Bird song learning causes and consequences" in Ethology, Ecology & Evolution, Vol 1 No 1, pp.19-46.

Slater, P. J. B. & S. A. Ince (1979), "Cultural evolution in chaffinch song" in Behaviour, Vol 71, pp147-166.

Smith, J. Maynard (1964) "Group selection and Kin selection" in Nature Vol 201, pp1147-1147 (1966), "Sympatric speciation" in The American Naturalist Vol 100 No 916, pp 637-630 (1974), "The theory of Games and the evolution of animal conflicts" in Journal of Theoretical Biology Vol 47, pp209-221. (1979), "Game theory and the evolution of behavior" in Proceedings of the Royal Society of London, B Vol 205, pp 475-488. & D Harper (1988), "The evolution of aggression: Can selection generate variability" in Philosophical Transactions of Royal Society of London B vol. 319 pp.557-570.

Sorenson, Micheal D., Mark E Hauber & Scott R Derrickson (2010), "Sexual imprinting misguides species recognition in a facultative interspecific brood parasite' in Proceeding of the Royal Society of Biological Science Vol 277, pp.3079-3085.

Stoddard, P K., M. D. Beecher, S. E. Campbell & C.Horning (1992), "Song type matching in the song sparrow" in Canadian Journal of Zoology vol 70, pp.1440-1444.

Sullivan, Brian K (2009), "Mate recognition, species boundaries and fallacy of 'species recognition'" in The Open Zoology Journal, Vol. 2, pp. 86-90.

Takahashi, Mariko, Hiroyuki Arita, Mariko Hiraiwa-Hasegawa, & Toshikazu Hasegawa (2007) "Peahens do not prefer peacocks with more elaborate traits" in Animal Behaviour Vol 75, pp 1209-1219.

Tchernichovskil, Ofer, Partha P. Mitra, Thierry Lints, & Fernando Nottebohm (2001),"Dynamics of the vocal imitation process: how a zebra finch learns its song" in Science, Vol 291, no 5513,pp2564-2569,

Templeton, Alan R (1981), "Mechanims of speciation—A population genetic approach" in Annual Review of Ecology and Systematics vol 12, pp23-48. (1980) "Modes of speciation and inferences based on genetic distance" in Evolution Vol 14 no 4, pp719-729.

Thorpe, W H (1958) "The learning of song patterns by birds with especial reference to the song of the chaffinch fringalla coelebs" in Ibis, Vol 100, No 4, pp.535-570.

Trivers, R L & H Hare (1975), "Haplodiploidy and the evolution of the social insects" in Science Vol 179, pp. 90-92.

Verrell, Paul A. (1988), "Stabilizing selection, sexual selection and speciation: A view of specific-mate recognition systems" in Systematic Zoology Vol 37, No 2, June, pp. 209-215.

Wallace, Alfred Russell (1867), in Proceedings of the Entomological Society of London, 4 March, pp. IXXX-IXXXi.

Yoktan, Kinneret. Eli Geffen, Amiyaal Hani *et al.*, (2011) "Vocal dialect and genetic subdivision along a geographic gradient in the orange-tufted sunbird" in Behavioral Ecology and Sociobiology, Vol. 65, pp. 1389- 1402.

Zahavi, Amotz (1975) "Mate selection - A selection for a handicap" in Journal of Theoretical Biology Vol 53, pp205-214. (1995) "Altruism as a handicap :The limitations of Kin Selection and Reciprocity' in Journal of Avian Biology Vol 26, No 1 pp. 1-3.

Zuk, Marlene, Kristine Johnson, Randy Thornhill, & J David Ligon *et al.*, (1990), "Female mate choice among red jungle fowl" in Evolution Vol 44, No 3(Mar), pp. 477-485.

www.ingramcontent.com/pod-product-compliance
Lightning Source LLC
Chambersburg PA
CBHW050516190326
41458CB00005B/1558